DATE DUE

~~AP 2 7 '98~~			
~~NV 1 3 '98~~			

DEMCO 38-296

Springer Series in Statistics

Advisors:
P. Bickel, P. Diggle, S. Fienberg, K. Krickeberg,
I. Olkin, N. Wermuth, S. Zeger

Springer
New York
Berlin
Heidelberg
Barcelona
Budapest
Hong Kong
London
Milan
Paris
Santa Clara
Singapore
Tokyo

Springer Series in Statistics

(continued after index)

Sylvie Huet
Annie Bouvier
Marie-Anne Gruet
Emmanuel Jolivet

Statistical Tools for Nonlinear Regression

A Practical Guide
with S-PLUS Examples

With 45 Illustrations

 Springer

Statistical tools for
nonlinear regression

...rie-Anne Gruet
...RA Laboratoire de Biométrie,
...52 Jouy-en-Josas Cedex,
...nce

Anne Bouvier
INRA Laboratoire de Biométrie,
78352 Jouy-en-Josas Cedex,
France

Emmanuel Jolivet
INRA SESAMES,
147 Rue de l'Université,
75338 Paris Cedex 07,
France

Mathematics Subject Classification (1991): 02.70, 62F03, 62F25, 62J02, 62J20

Library of Congress Cataloging-in-Publication Data
Statistical tools for nonlinear regression : a practical guide / S.
  Huet . . . [et al.].
      p.   cm. — (Springer series in statistics)
    Includes bibliographical references and index.
    ISBN 0-387-94727-2 (hard : alk. paper)
    1. Regression analysis.   2. Nonlinear theories.   3. Parameter
estimation.   I. Huet, S. (Sylvie)   II. Series.
  QA278.2.S73   1996
  519.5'36—dc20   96-13753

Printed on acid-free paper.

Production managed by Natalie Johnson; manufacturing supervised by Jeffrey Taub.
Camera-ready copy prepared using the authors' LaTeX files.
Printed and bound by Braun Brumfield, Inc., Ann Arbor, MI.
Printed in the United States of America.

9 8 7 6 5 4 3 2 1

ISBN 0-387-94727-2 Springer-Verlag New York Berlin Heidelberg   SPIN 10533039

Preface

If you need to analyze a data set using a parametric nonlinear regression model, if you are not on familiar terms with statistics and software, and if you make do with **S-PLUS**, this book is for you. In each chapter we start by presenting practical examples. We then describe the problems posed by these examples in terms of statistical problems, and we demonstrate how to solve these problems. Finally, we apply the proposed methods to the example data sets. You will not find any mathematical proofs here. Rather, we try when possible to explain the solutions using intuitive arguments. This is really a *cook book*.

Most of the methods proposed in the book are derived from classical nonlinear regression theory, but we have also made attempts to provide you with more modern methods that have proved to perform well in practice. Although the theoretical grounds are not developed here, we give, when appropriate, some technical background using a sans serif type style. You can skip these passages if you are not interested in this information.

The first chapter introduces several examples, from experiments in agronomy and biochemistry, to which we will return throughout the book. Each example illustrates a different problem, and we show how to methodically handle these problems by using parametric nonlinear regression models. Because the term *parametric model* means that all of the information in the experiments is assumed to be contained in the parameters occuring in the model, we first demonstrate, in chapter 1, how to estimate the parameters. In chapter 2 we describe how to determine the accuracy of the estimators. Chapter 3 introduces some new examples and presents methods for handling nonlinear regression models when the variances are heterogeneous with few or no replications. In chapter 4 we demonstrate methods for checking if the assumptions on which the statistical analysis is based are accurate, and we provide methods for detecting and correcting any misspecification that might exist. In chapter 5 we describe how to calculate prediction and calibration confidence intervals.

Because good software is necessary for handling nonlinear regression data, we provide, at the end of each chapter, a step-by-step description of how to treat our examples using **nls2** [BH94], the software we have used throughout this book. **nls2** is a software implemented as an extension of the statistical system **S-PLUS**, available by http://www-bia.inra.fr/ or by ftp www-bia.inra.fr in pub/log/nls2, and offers the capability of implementing all of the methods presented in this book.

Last but not least, we are grateful to Suzie Zweizig for a careful rereading of our English. Thanks to her, we hope that you find this book readable!

Contents

1

Nonlinear regression model and parameter estimation

In this chapter we describe five examples of experiments in agronomy and biochemistry; each illustrates a different problem. Because all of the information in the experiments is assumed to be contained in a set of parameters, we will first describe, in this chapter, the parametric nonlinear regression model and how to estimate the parameters of this model.[1]

1.1 Examples

1.1.1 Pasture regrowth: estimating a growth curve

In biology, the growth curve is of great interest. In this example (treated by Ratkowsky [Rat83]), we observe the yield of pasture regrowth versus time since the last grazing. The results are reported in table 1.1 and figure 1.1. The following model is assumed: the yield at time x_i, Y_i, is written as $Y_i = f(x_i, \theta) + \varepsilon_i$, where $f(x, \theta)$ describes the relationship between the yield and the time x. The errors ε_i are centered random variables ($E(\varepsilon_i) = 0$). Because the measurements are done on different experimental units, the errors can be assumed to be independent. Moreover, to complete the regression model, we will assume that the variance of ε_i exists and equals σ^2.

---

[1]We use this sans serif type style when we give some technical background. You can skip these passages if you are not interested in this information.

TABLE 1.1. Data of yield of pasture regrowth versus time (the units are not mentioned)

Time after pasture	9	14	21	28	42
Yield	8.93	10.8	18.59	22.33	39.35
Time after pasture	57	63	70	79	
Yield	56.11	61.73	64.62	67.08	

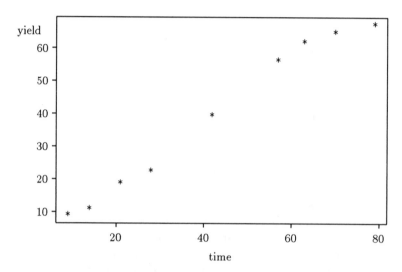

FIGURE 1.1. Pasture regrowth example: observed responses versus time

The model under study is the following:

$$\left. \begin{array}{rcl} Y_i & = & f(x_i, \theta) + \varepsilon_i \\ \mathrm{Var}(\varepsilon_i) & = & \sigma^2, \ \mathrm{E}(\varepsilon_i) = 0 \end{array} \right\}. \tag{1.1}$$

In this example, a possible choice for f is the Weibull model:

$$f(x, \theta) = \theta_1 - \theta_2 \exp\left(-\exp(\theta_3 + \theta_4 \log x)\right). \tag{1.2}$$

Depending on the value of θ_4, f presents an inflexion point, or an exponential increase; see figure 1.2.

Generally the choice of the function f depends on knowing the observed biological phenomena. Nevertheless, in this example, figure 1.1 indicates that the response curve may present an inflexion point, making the Weibull model a good choice.

Obviously, it is important to choose the function f well. In this chapter we will show how to estimate the parameters assuming that f is correct. In chapter 4 we will describe how to validate the choice of f.

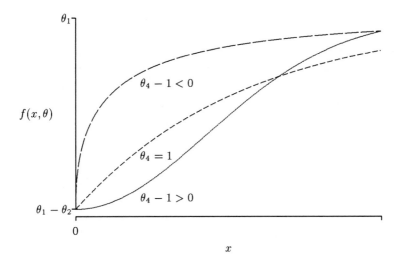

FIGURE 1.2. Growth models described with the Weibull function:
$f(x, \theta) = \theta_1 - \theta_2 \exp(-e^{\theta_3} x^{\theta_4})$

1.1.2 Radioimmunological assay of cortisol: estimating a calibration curve

Because the amount of hormone contained in a preparation cannot be measured directly, a two-step process is necessary to estimate an unknown dose of hormone: first we must establish *a calibration curve*, and then we invert the calibration curve to find the dose of the hormone.

The calibration curve is estimated by using a radioimmunological assay (or RIA). This assay is based on the fact that the hormone H and its marked isotope H* behave similarly with respect to their specific antibody A. Thus, for fixed quantities of antibody [A] and radioactive hormone [H*], the quantity of linked complex [AH*] decreases when the quantity of cold hormone [H] increases. This property establishes a *calibration curve*. For known dilutions of a purified hormone, the responses [AH*] are measured in terms of c.p.m. (counts per minute). Because the relation between the quantity of linked complex [AH*] and the dose of hormone H varies with the experimental conditions, the calibration curve is evaluated for each assay.

Then, for a preparation containing an unknown quantity of the hormone H, the response [AH*] is measured. The unknown dose is calculated by inverting the calibration curve.

The data for the calibration curve of an RIA of cortisol are given in table 1.2 (data from the Laboratoire de Physiologie de la Lactation, INRA). Figure 1.3 shows the observed responses as functions of the logarithm of the dose, which is the usual transformation for the x-axis. In this experiment, the response has been observed for the zero dose and the infinite dose. These doses are represented in figure 1.3 by the values -3 and +2, respectively.

TABLE 1.2. Data for the calibration curve of an RIA of cortisol

Dose in ng/.1 ml	Response in c.p.m.			
0	2868	2785	2849	2805
0	2779	2588	2701	2752
0.02	2615	2651	2506	2498
0.04	2474	2573	2378	2494
0.06	2152	2307	2101	2216
0.08	2114	2052	2016	2030
0.1	1862	1935	1800	1871
0.2	1364	1412	1377	1304
0.4	910	919	855	875
0.6	702	701	689	696
0.8	586	596	561	562
1	501	495	478	493
1.5	392	358	399	394
2	330	351	343	333
4	250	261	244	242
∞	131	135	134	133

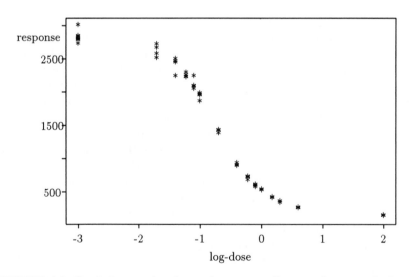

FIGURE 1.3. Cortisol example: observed responses (in c.p.m.) versus the logarithms in base 10 of the dose

TABLE 1.3. Cortisol example: mean and empirical variance of the response for each value of the dose

Dose	0	0.02	0.04	0.06	0.08
Means: $Y_{i\bullet}$	2765.875	2567.5	2479.75	2194	2053
Variances: s_i^2	6931	4460	4821	5916	1405

Dose	0.1	0.2	0.4	0.6	0.8
Means : $Y_{i\bullet}$	1867	1364.25	889.75	697	576.25
Variances: s_i^2	2288	1518	673	26	230

Dose	1	1.5	2	4	∞
Means: $Y_{i\bullet}$	491.75	385.75	339.25	249.25	133.25
Variances: s_i^2	72	263	69	55	2

The model generally used to describe the variations of the response Y versus the log-dose x is the Richards function (the generalized logistic function). This model depends on 5 parameters: θ_2, the upper asymptote, is the response for a null dose of hormone H; θ_1, the lower asymptote, is the response for a theoretical infinite dose of H; the other three parameters describe the shape of the decrease of the curve. The equation of the calibration curve is

$$f(x,\theta) = \theta_1 + \frac{\theta_2 - \theta_1}{(1 + \exp(\theta_3 + \theta_4 x))^{\theta_5}}. \tag{1.3}$$

Once we describe the trend of the variations of Y versus x, we also have to consider the errors $\varepsilon = Y - f(x,\theta)$. Looking carefully at figure 1.3, we see that the variability of Y depends on the level of the response. For each value of x, with the response Y being observed with replications, the model can be written in the following manner: $Y_{ij} = f(x_i,\theta) + \varepsilon_{ij}$, j varying from 1 to n_i (n_i equals 8 or 4) and i from 1 to k ($k = 15$). For each value of i the empirical variance of the response can be calculated:

$$\left. \begin{array}{l} s_i^2 = \frac{1}{n_i} \sum_{j=1}^{n_i} (Y_{ij} - Y_{i\bullet})^2 \\ \text{with} \quad Y_{i\bullet} = \frac{1}{n_i} \sum_{j=1}^{k} Y_{ij} \end{array} \right\}. \tag{1.4}$$

Table 1.3 gives the values of s_i^2 and $Y_{i\bullet}$. Since the variance of the response is clearly heterogeneous, it will be assumed that $\text{Var}(\varepsilon_{ij}) = \sigma_i^2$. The model is the following:

$$\left. \begin{array}{rcl} Y_{ij} & = & f(x_i,\theta) + \varepsilon_{ij} \\ \text{Var}(\varepsilon_{ij}) & = & \sigma_i^2, \ \text{E}(\varepsilon_{ij}) = 0 \end{array} \right\}, \tag{1.5}$$

where $j = 1, \ldots n_i$, $i = 1, \ldots k$, and the total number of observations equals $n = \sum_{i=1}^{k} n_i$. f is defined by equation (1.3), and the ε_{ij} are independent centered random variables. In this experiment, however, we can be even more precise about the heteroscedasticity, that is, about the variations of σ_i^2, because the observations are counts that can safely be assumed to be distributed as Poisson variables, for which variance equals expectation. A reasonable model for the variance could be $\sigma_i^2 = f(x_i, \theta)$, which will be generalized to $\sigma_i^2 = \sigma^2 f(x_i, \theta)^\tau$, where τ can be known or estimated (see chapter 3).

The aim of this experiment is to estimate an unknown dose of hormone using the calibration curve. Let μ be the expected value of the response for a preparation where the dose of hormone D (or its logarithm X) is unknown; then X is the inverse of f calculated in μ (if μ lies strictly between θ_1 and θ_2):

$$X = f^{-1}(\mu, \theta)$$

$$X = \frac{1}{\theta_4}\left\{\log\left[\exp\frac{1}{\theta_5}\log\frac{\theta_2 - \theta_1}{\mu - \theta_1} - 1\right] - \theta_3\right\}. \tag{1.6}$$

We will discuss how to estimate the calibration curve and X in this chapter. The problem of calculating the accuracy of the estimation of X will be treated in chapter 2. The calibration problem will be treated in detail in chapter 5.

1.1.3 Antibodies anticoronavirus assayed by an ELISA test: comparing several response curves

This experiment uses an ELISA test to detect the presence of anticoronavirus antibodies in the serum of calves and cows. It shows how we can estimate the parameters of two curves together so that we can ultimately determine if the curves are identical up to a horizontal shift. The complete description of the experiment is presented in [HLV87]. Here we limit the problem to the comparison of antibody levels in two serum samples taken in May and June for one cow.

An ELISA test is a collection of observed optical densities, Y, for different serum dilutions, say d. The results are reported in table 1.4 and figure 1.4.

Assuming that the serums are assayed under the same experimental conditions, the problem is to quantify the differences (in terms of antibody level) between the serums using their ELISA response curves. For that purpose, the biological assay techniques described in [Fin78] can be used to estimate the *potency* of one serum *relative* to another serum. The potency ρ is defined in the following manner: one unit of serum taken in May is assumed to produce the same response as ρ units of serum taken in June. This means that the two serums must contain the same effective constituent, the antibody, and that all other constituents are without effect on the response. Hence one preparation behaves as a dilution of the other in an inert diluent.

TABLE 1.4. ELISA example: observed values of Y for the serums taken in May and June. Two values of the response are observed

dilution	Response (optical density)			
d	May		June	
1/30	1.909	1.956	1.886	1.880
1/90	1.856	1.876	1.853	1.870
1/270	1.838	1.841	1.747	1.772
1/810	1.579	1.584	1.424	1.406
1/2430	1.057	1.072	0.781	0.759
1/7290	0.566	0.561	0.377	0.376
1/21869	0.225	0.229	0.153	0.138
1/65609	0.072	0.114	0.053	0.058

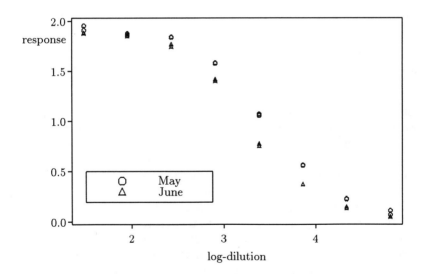

FIGURE 1.4. ELISA example: observed responses versus the logarithms of the dilution, $x = -\log_{10} d$

Let $F^{\text{May}}(d)$ and $F^{\text{June}}(d)$ be the response functions for the serums taken in May and June, respectively. If the assumptions defined above are fulfilled, then the two regression functions must be related in the following manner: $F^{\text{May}}(d) = F^{\text{June}}(\rho d)$.

The relationship between the optical density Y and the logarithm of the dilution $x = \log_{10}(1/d)$ is usually modeled by a sigmoidal curve: $F^{\text{May}}(d) = f(x, \theta^{\text{May}})$ and $F^{\text{June}}(d) = f(x, \theta^{\text{June}})$, with

$$f(x, \theta) = \theta_1 + \frac{\theta_2 - \theta_1}{1 + \exp \theta_3 (x - \theta_4)}.$$

Thus, the estimation of ρ is based on an assumption of *parallelism* between the two response curves:

$$f(x, \theta^{\text{May}}) = f(x - \beta; \theta^{\text{June}}), \tag{1.7}$$

where $\beta = \log_{10}(1/\rho)$.

In this chapter we will estimate the parameters of the regression functions. Obviously, one must also estimate β. For that purpose, the two curves must be compared to see if they verify equation (1.7). This procedure will be demonstrated in chapter 2.

1.1.4 Comparison of immature and mature goat ovocytes; comparing parameters

This experiment comparing the responses of immature and mature goat ovocytes to an hyper-osmotic test demonstrates how we can estimate the parameters of two different data sets for the ultimate purpose of determining the differences between these sets of parameters.

Cellular damage that results from exposure to low temperature can be circumvented by the use of permeable organic solvents acting as cryoprotectants. Cell water permeability is higher than cell cryoprotectant permeability. Thus, when cells are exposed to these solvents, they lose water and shrink in response to the variation of osmotic pressure that is created between the intra- and extracellular compartments. Then, as the compound permeates, water reenters the cell, and the cell reexpands until the system reaches an osmotic equilibrium. The results obtained using immature and ovulated (mature) ovocytes exposed to propane-diol, a permeable compound, are presented in [LGR94].

The cell volume during equilibration is recorded at each time t. In order to obtain water (P_w) and propane-diol (P_S) permeabilities, the following equations are used:

$$
\left.
\begin{array}{rcl}
\dfrac{V(t)}{V_0} &=& \dfrac{1}{V_0}\left(V_w(t) + V_s(t) + V_x\right), \\[2mm]
\dfrac{dV_w}{dt} &=& P_w\left(\gamma_1 + \gamma_2\dfrac{1}{V_w} + \gamma_3\dfrac{V_s}{V_w}\right) \\[2mm]
\dfrac{dV_s}{dt} &=& P_s\left(\gamma_4 + \gamma_5\dfrac{V_s}{V_w}\right)
\end{array}
\right\}. \tag{1.8}
$$

The first equation expresses the cell volume $V(t)$ as a percentage of initial isotonic cell volume V_0 at any given time t during equilibration; V_x is a known parameter depending on the stage of the oocyte (mature or immature); $V_w(t)$ and $V_s(t)$ are the volumes of intracellular water and cryoprotectant, respectively; and the parameters γ are known. The data are reported in table 1.5.

In this example, the variations of $Y = V(t)/V_0$ versus the time t are approximated by

$$
f(t, P_w, P_s) = \frac{1}{V_0}\left(V_w(t) + V_s(t) + V_x\right),
$$

where V_w and V_s are the solutions of a system of ordinary differential equations depending on P_w and P_s.

Our aim here is to estimate the parameters P_w and P_s for both types of ovocytes and to compare their values. We also are interested in the temporal evolution of intracellular propane-diol penetration ($V_s(t)$), expressed as the fraction of initial isotonic cell volume V_0.

1.1.5 Isomerization: more than one independent variable

This data, given by Carr [Car60] and extensively treated by Bates and Watts [BW88], illustrates how to estimate parameters when there is more than one variable.

The reaction rate of the catalytic isomerization of n-pentane to isopentane depends on various factors such as the partial pressures of the needed products (employed to speed up the reaction). The differential reaction rate is expressed as grams of iso-pentane produced per gram of catalyst per hour, and the instantaneous partial pressure of each component is measured. The data are reproduced in table 1.6.

A common form of the model is the following:

$$
f(x, \theta) = \frac{\theta_1\theta_3(P - I/1.632)}{1 + \theta_2 H + \theta_3 P + \theta_4 I}, \tag{1.9}
$$

where x is a three-dimensional variate, $x = (H, P, I)$, where H, P, and I are the partial pressures of hydrogen, n-pentane and iso-pentane. In this example we will estimate the θ. In chapter 2 we will calculate the confidence intervals.

TABLE 1.5. Fraction of volume at time t for two stages (mature and immature) of ovocytes

Times	Fraction of cell volume for mature ovocytes						
0.5	0.6833	0.6870	0.7553	0.6012	0.6655	0.7630	0.7380
1	0.6275	0.5291	0.5837	0.5425	0.5775	0.6640	0.6022
1.5	0.6743	0.5890	0.5837	0.5713	0.6309	0.6960	0.6344
2	0.7290	0.6205	0.5912	0.5936	0.6932	0.7544	0.6849
2.5	0.7479	0.6532	0.6064	0.6164	0.7132	0.7717	0.7200
3	0.7672	0.6870	0.6376	0.6884	0.7494	0.8070	0.7569
4	0.7865	0.7219	0.6456	0.7747	0.8064	0.8342	0.8100
5	0.8265	0.7580	0.6782	0.8205	0.8554	0.8527	0.8424
7	0.8897	0.8346	0.8238	0.8680	0.8955	0.8621	0.8490
10	0.9250	0.9000	0.9622	0.9477	0.9321	0.8809	0.8725
12	0.9480	0.9290	1.0000	0.9700	0.9362	0.8905	0.8930
15	0.9820	0.9550	1.0000	1.0000	0.9784	0.9100	0.9242
20	1.0000	0.9800	1.0000	1.0000	1.0000	0.9592	0.9779
Times	Fraction of cell volume for immature ovocytes						
0.5	0.4536	0.4690	0.6622	0.621	0.513		
1	0.4297	0.4552	0.5530	0.370	0.425		
1.5	0.4876	0.4690	0.5530	0.400	0.450		
2	0.5336	0.4760	0.5535	0.420	0.475		
2.5	0.5536	0.4830	0.5881	0.460	0.500		
3	0.5618	0.5047	0.6097	0.490	0.550		
3.5	0.6065	0.5269	0.6319	0.520	0.600		
4	0.6365	0.5421	0.6699	0.550	0.650		
4.5	0.6990	0.5814	0.6935	0.570	0.700		
5	0.7533	0.6225	0.7176	0.580	0.750		
6.5	0.7823	0.6655	0.7422	0.630	0.790		
8	0.8477	0.7105	0.7674	0.690	0.830		
10	0.8966	0.8265	0.8464	0.750	0.880		
12	0.9578	0.9105	0.9115	0.800	0.950		
15	1.0000	0.9658	0.9404	0.900	1.000		
20	1.0000	1.0000	0.9800	0.941	1.000		

TABLE 1.6. Reaction rate of the catalytic isomerization of n-pentane to iso-pentane versus the partial pressures of hydrogen, n-pentane, and iso-pentane

Partial pressure of			Reaction
Hydrogen	n-Pentane	Iso-pentane	rate
205.8	90.9	37.1	3.541
404.8	92.9	36.3	2.397
209.7	174.9	49.4	6.694
401.6	187.2	44.9	4.722
224.9	92.7	116.3	0.593
402.6	102.2	128.9	0.268
212.7	186.9	134.4	2.797
406.2	192.6	134.9	2.451
133.3	140.8	87.6	3.196
470.9	144.2	86.9	2.021
300.0	68.3	81.7	0.896
301.6	214.6	101.7	5.084
297.3	142.2	10.5	5.686
314.0	146.7	157.1	1.193
305.7	142.0	86.0	2.648
300.1	143.7	90.2	3.303
305.4	141.1	87.4	3.054
305.2	141.5	87.0	3.302
300.1	83.0	66.4	1.271
106.6	209.6	33.0	11.648
417.2	83.9	32.9	2.002
251.0	294.4	41.5	9.604
250.3	148.0	14.7	7.754
145.1	291.0	50.2	11.59

1.2   The parametric nonlinear regression model

Let us introduce the following notation for defining the parametric non-linear regression model: n_i replications of the response Y are observed for each value of the independent variable x_i; let Y_{ij}, for $j = 1, \ldots n_i$, be these observations. i varies from 1 to k, and the total number of observations is $n = \sum_{i=1}^{k} n_i$. In example 1.1.1, n_i equals 1 for all values of i, and $k = n$. In example 1.1.5, the variable x is of dimension three: x_i is a vector of values taken by (H, P, I).

It is assumed that the true regression relationship between Y and x is the sum of a systematic part, described by a function $\mu(x)$, and a random part. Generally, the function $\mu(x)$ is unknown and is approximated by a parametric function f, called the regression function, that depends on unknown parameters θ:

$$Y_{ij} = f(x_i, \theta) + \varepsilon_{ij}.$$

θ is a vector of p parameters $\theta_1, \theta_2, \ldots \theta_p$. The function f does not need to be known explicitly; in example 1.1.4, f is a function of the solution of differential equations.

ε is a random error equal, by construction, to the discrepancy between Y and $f(x, \theta)$. Let σ_i^2 be the variance of ε_{ij}. The values of σ_i^2, or their variations as functions of x_i, are unknown and must be approximated. In some situations the difference in the variances, $\sigma_i - \sigma_{i+1}$, is small. Thus, there is a high confidence in the approximate homogeneity of the variances, and we can then assume that $\mathrm{Var}(\varepsilon_{ij}) = \sigma^2$. In other cases, because of the nature of the observed response or because of the aspect of the data on a graph, there is evidence against the assumption of homogeneous errors. In these cases, the true (unknown) variations of σ_i^2 are approximated by a function g called the variance function such that $\mathrm{Var}(\varepsilon_{ij}) = g(x_i, \sigma^2, \theta, \tau)$. In most cases, g will be assumed to depend on f; for example, $g(x, \sigma^2, \theta, \tau) = \sigma^2 f(x, \theta)^\tau$, where τ is a set of parameters that can be assumed to be known or that have to be estimated.

We simplify the necessary technical assumptions by assuming that each θ_a, $a = 1, \ldots p$, varies in the interior of an interval. The function f is assumed to be twice continuously differentiable with respect to the parameters θ.

1.3   estima-1

Since the problem is to estimate the unknown vector θ, a natural solution is to choose the value of θ that minimizes the distances between the values of $f(x, \theta)$ and the observations of Y. For example, one can choose the value of θ that minimizes the sum of squares $C(\theta)$ defined by

$$C(\theta) = \sum_{i=1}^{k}\sum_{j=1}^{n_i}(Y_{ij}-f(x_i,\theta))^2. \tag{1.10}$$

Let $\widehat{\theta}$ be this value. $\widehat{\theta}$ is the least squares estimator of θ. If we assume that $\mathrm{Var}(\varepsilon_{ij})=\sigma^2$, an estimate of σ^2 is obtained as

$$\widehat{\sigma}^2 = \frac{C(\widehat{\theta})}{n}. \tag{1.11}$$

Under the assumptions stated in the preceding paragraph, $\widehat{\theta}$ is also the solution of the set of p equations:

$$\sum_{i=1}^{k}\frac{\partial f}{\partial\theta_a}(x_i,\theta)\sum_{j=1}^{n_i}(Y_{ij}-f(x_i,\theta)) = 0,$$

for $a=1,\ldots p$, where $\dfrac{\partial f}{\partial\theta_a}$ is the partial derivative of f with respect to θ_a.

Because f is nonlinear in θ, no explicit solution can be calculated. Instead an iterative procedure is needed.

In the case of heterogeneous variances $(\mathrm{Var}(\varepsilon_{ij})=\sigma_i^2)$, it is natural to favor observations with small variances by weighting the sum of squares. The σ_i^2 are unknown, so they must be replaced by an estimate. If each n_i is big enough, say 4, as in example 1.1.2, then σ_i^2 can be estimated by s_i^2, the empirical variance (see equation (1.4)). The weighted sum of squares is

$$W(\theta) = \sum_{i=1}^{k}\sum_{j=1}^{n_i}\frac{(Y_{ij}-f(x_i,\theta))^2}{s_i^2}, \tag{1.12}$$

and the value of θ that minimizes $W(\theta)$ is the weighted least squares estimator of θ. The problem of estimating the parameters in heteroscedastic models is treated more generally in chapter 3.

1.4 Applications

1.4.1 Pasture regrowth: parameter estimation and graph of observed and adjusted response values

Model. The regression function is

$$f(x,\theta) = \theta_1 - \theta_2\exp\left(-\exp(\theta_3+\theta_4\log x)\right),$$

and the variances are homogeneous: $\mathrm{Var}(\varepsilon_i)=\sigma^2$.

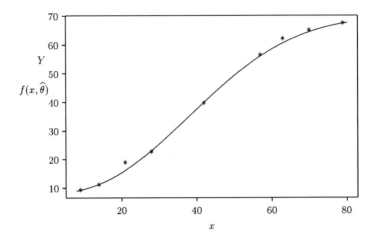

FIGURE 1.5. Pasture regrowth example: graph of observed and adjusted response values

Method. The parameters are estimated by minimizing the sum of squares $C(\theta)$; see equation (1.10).

Results.

Parameters	Estimated values
θ_1	69.95
θ_2	61.68
θ_3	-9.209
θ_4	2.378

The adjusted response curve, $f(x, \widehat{\theta})$, is shown in figure 1.5. It is defined by the following equation:

$$f(x, \widehat{\theta}) = 69.95 - 61.68 \exp\left(-\exp(-9.209 + 2.378 \log x)\right).$$

1.4.2 Cortisol assay: parameter estimation and graph of observed and adjusted response values

Model. The regression function is

$$f(x, \theta) = \theta_1 + \frac{\theta_2 - \theta_1}{(1 + \exp(\theta_3 + \theta_4 x))^{\theta_5}},$$

and the variances are heterogeneous: $\mathrm{Var}(\varepsilon_i) = \sigma_i^2$.

Method. The parameters are estimated by minimizing the weighted sum of squares $W(\theta)$; see equation (1.12).

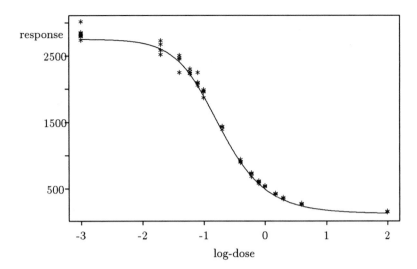

FIGURE 1.6. Cortisol assay example: graph of observed and adjusted response values

Results.

Parameters	Estimated values
θ_1	133.30
θ_2	2759.8
θ_3	3.0057
θ_4	3.1497
θ_5	0.64309

The adjusted response curve, $f(x, \widehat{\theta})$, is shown in figure 1.6. It is defined by the following equation:

$$f(x, \widehat{\theta}) = 133.3 + \frac{2759.8 - 133.3}{(1 + \exp(3.0057 + 3.1497x))^{0.64309}}.$$

1.4.3 ELISA test: parameter estimation and graph of observed and adjusted curves for May and June

Model. The regression function is

$$f(x, \theta) = \theta_1 + \frac{\theta_2 - \theta_1}{1 + \exp \theta_3 \left(x - \theta_4\right)},$$

and the variances are homogeneous: $\text{Var}(\varepsilon_i) = \sigma^2$.

Method. The parameters are estimated by minimizing the sum of squares $C(\theta)$; see equation (1.10).

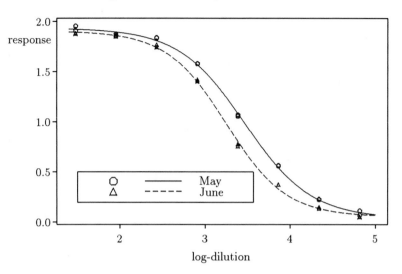

FIGURE 1.7. ELISA test example: graph of observed and adjusted curves for May and June

Results.

Parameters	Estimated values
θ_1^{May}	0.04279
θ_2^{May}	1.936
θ_3^{May}	2.568
θ_4^{May}	3.467
θ_1^{June}	0.0581
θ_2^{June}	1.909
θ_3^{June}	2.836
θ_4^{June}	3.251

The adjusted response curves for May and June are shown in figure 1.7.

1.4.4 Ovocytes: parameter estimation and graph of observed and adjusted volume of mature and immature ovocytes in propane-diol

Model. The regression function is

$$f(t, P_w, P_s) = \frac{1}{V_0}(V_w(t) + V_s(t) + V_x),$$

where V_w and V_s are solutions of equation (1.8). The variances are homogeneous: $\text{Var}(\varepsilon_i) = \sigma^2$.

Method. The parameters are estimated by minimizing the sum of squares $C(\theta)$; see equation (1.10).

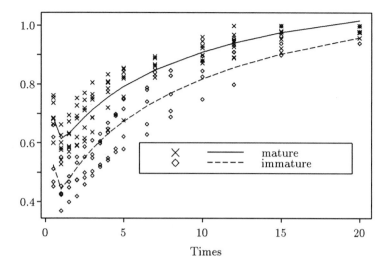

FIGURE 1.8. Ovocytes example: graph of observed and adjusted volume of mature and immature ovocytes in propane-diol

Results.

	Parameters	Estimated values
mature	P_w	0.0784
ovocytes	P_s	0.00147
immature	P_w	0.1093
ovocytes	P_s	0.00097

The adjusted response curves for mature and immature ovocytes are shown in figure 1.8.

1.4.5 Isomerization: parameter estimation and graph of adjusted versus observed values

Model. The regression function is

$$f(x, \theta) = \frac{\theta_1 \theta_3 (P - I/1.632)}{1 + \theta_2 H + \theta_3 P + \theta_4 I},$$

where x is a three-dimensional variate, $x = (H, P, I)$, and where the variances are homogeneous: $\mathrm{Var}(\varepsilon_i) = \sigma^2$.

Method. The parameters are estimated by minimizing the sum of squares $C(\theta)$; see equation (1.10).

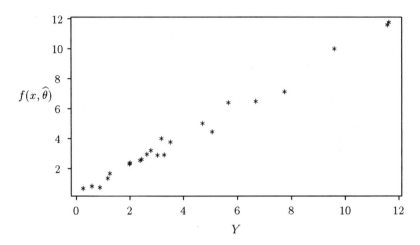

FIGURE 1.9. Isomerization example: graph of adjusted versus observed values

Results.

Parameters	Estimated values
θ_1	35.9191
θ_2	0.07086
θ_3	0.03774
θ_4	0.1672

Figure 1.9 shows the values of the adjusted response curve, $f(x_i, \widehat{\theta})$, versus the observations Y_i. $f(x, \widehat{\theta})$ is defined by the following equation:

$$f(x, \widehat{\theta}) = \frac{1.356(P - I/1.632)}{1 + 0.071H + 0.038P + 0.167I}.$$

1.5   Conclusion and references

For each example presented here, we chose a model for the response curve and calculated an estimate of the parameters, but we still need to assess the accuracy of our estimates. Consider the RIA example: we were able to calculate an estimate of X, the inverse of f calculated in a known value μ; however, now we need to calculate the accuracy of this estimate. For the ELISA data, we estimated two response curves, but we are interested in testing the hypothesis of parallelism between these curves. In example 1.1.4, we wanted to calculate confidence regions for the pair (P_w, P_s).

The next chapter describes the tools that help us answer these questions.

For more information, the interested reader can find a complete description of statistical and numerical methods in nonlinear regression models in the book of G. Seber and C. Wild [SW89]. See also the book of H. Bunke and

O. Bunke [BB89] and the book of A. Gallant [Gal87]. These are strong in
theory and cover mainly the case of homogeneous variances. Users of econo-
metrics should be interested in the third one, which covers the field of mul-
tivariate nonlinear regressions and dynamic nonlinear models. The books of
S. Huet et al. [HJM91], and R. Carroll and D. Ruppert [CR88] consider the
case of heterogeneous variances. Response curves that are currently used in
biological frameworks are described in the books of D. Ratkowsky [Rat89]
and J. Lebreton and C. Millier [LM82]. Other books are concerned with
specific subjects, such as the books of D. Bates and D. Watts [BW88]
and D. Ratkowsky [Rat83] about nonlinear curvatures, or the book of G.
Ross [Ros90] about parameter transformations. The book of W. Venables
and B. Ripley [VR94] shows how to analyze data using **S-PLUS**.

1.6   Using **nls2**

This section reproduces the commands and files used in this chapter to
treat the examples by using **nls2**. We assume that the user is familiar with
S-PLUS and has access to the help files of **nls2**, which provide all of the
necessary information about syntax and possibilities.

The outputs are not reproduced; the main results are shown in the preceding part.

Typographical conventions. We use the prompt $ for the operating system
commands and the prompt > for the **S-PLUS** commands. Continuation
lines are indented.

Pasture regrowth example

Creating the data

The experimental data (table 1.1, page 2) are stored in a *data-frame* structure:

```
$ Splus
> pasture <- data.frame(
    time=c(9, 14, 21, 28, 42, 57, 63, 70, 79),
    yield= c(8.93, 10.8, 18.59, 22.33, 39.35,
            56.11, 61.73, 64.62, 67.08))
```

Plot of the observed yield versus time

We plot the observed values of yield versus time by using the function
pldnls2:

```
> library("nls2") # attach the nls2 library
> X11()           # open a graphical device
> pldnls2(pasture, response.name="yield", X.name="time",
```

```
title = "Pasture regrowth example",
sub = "Observed response")
```

(See figure 1.1, page 2).

Description of the model

The model defined in section 1.4.1, page 13, is described using a symbolic syntax in a file called **pasture.mod1**:

```
resp yield;
varind time;
parresp p1, p2, p3, p4;
subroutine;
begin
yield=p1-p2*exp(-exp(p3+p4*log(time)));
end
```

Parameter estimation

Parameters are estimated by using the function **nls2**.

Before calling **nls2**, it is necessary to load the programs of the system **nls2** into the **S-PLUS** session; this is done by the function **loadnls2**.

The arguments of **nls2** are the name of the *data-frame*, the name of the file that describes the model and information about the statistical context, which includes the starting values of the parameters and possibly some other things, as the user chooses. Here we include, for example, the maximum number of iterations:

```
> loadnls2() # load the programs
> pasture.nl1 <- nls2(pasture, "pasture.mod1",
                list(theta.start= c(70, 60, 0, 1), max.iters=100))
>   # Print the estimated values of the parameters
> cat( "Estimated values of the parameters:\n ")
> print( pasture.nl1$theta); cat( "\n\n")
```

(Results are given in section 1.4.1, page 13.)

Plot of the observed and fitted yields versus time

We plot the observed and fitted response values versus time by using the function **plfit**:

```
> plfit(pasture.nl1, title = "Pasture regrowth example",
        sub = "Observed and fitted response")
```

(See figure 1.5, page 14).

Cortisol assay example

Creating the data

The experimental data (see table 1.2, page 4) are stored in a *data-frame*. We set the value of the zero dose to 0 and the value of the infinity dose to

10.

```
> corti <- data.frame(dose=c(
    rep(0,8),    rep(0.02,4), rep(0.04,4), rep(0.06,4),
    rep(0.08,4),rep(0.1,4),  rep(0.2,4),  rep(0.4,4),
    rep(0.6,4), rep(0.8,4),  rep(1,4),    rep(1.5,4),
    rep(2,4),   rep(4,4),    rep(10,4)),
                       cpm=c(
    2868,2785,2849,2805,2779,2588,2701,2752,
    2615,2651,2506,2498,2474,2573,2378,2494,
    2152,2307,2101,2216,2114,2052,2016,2030,
    1862,1935,1800,1871,1364,1412,1377,1304,
    910,919,855,875,702,701,689,696,
    586,596,561,562,501,495,478,493,
    392,358,399,394,330,351,343,333,
    250,261,244,242,131,135,134, 133))
```

Plot of the observed responses versus the logarithm of the dose

We plot the observed responses versus the logarithm of the dose by using the graphical function `plot` of **S-PLUS**:

```
> logdose<-c(rep(-3,8),log10(corti$dose[9:60]),rep(2,4))
> plot(logdose,corti$cpm,xlab="log-dose",ylab="response",
      title="Cortisol example",sub="Observed cpm")
```

(See figure 1.3, page 4).

Description of the model

The model, defined in section 1.4.2, page 14, is described in a file called `corti.mod1`.

The parameters `minf` and `pinf`, introduced by the key word `pbisresp`, are the minimal and maximal values accepted for the dose. They do not need to be estimated; they will be called *second-level parameters*.

```
resp cpm;
varind dose;
parresp n,d,a,b,g;
pbisresp minf,pinf;
subroutine;
begin
cpm= if dose <= minf then d else
        if dose >= pinf then n else
      n+(d-n)*exp(-g*log(1+exp(a+b*log10(dose))))
    fi fi;
end
```

Parameter estimation

The function `nls2` is used to estimate the parameters. Its first argument is the name of the *data-frame*, its second argument describes the model and its third one contains the starting values of the parameters.

The second argument, which describes the model, includes the name of the description file, the values of the *second-level parameters*, and information about the variance (VI means "Variance Intra-replications").

```
> corti.nl1<-nls2(corti,
    list(file="corti.mod1", gamf=c(0,10),vari.type="VI"),
    c(3000,30,0,1,1))
>    # Print the estimated values of the parameters
> cat( "Estimated values of the parameters:\n ")
> print( corti.nl1$theta); cat( "\n\n")
```

(Results are given in section 1.4.2, page 14.)

Plot of the observed and fitted response values

We plot the observed and fitted response values versus the logarithm of the dose by using graphical functions of **S-PLUS**:

```
> plot(logdose,corti$cpm,xlab="log-dose",ylab="response",
       title="Cortisol example",sub="Observed and fitted response")
> lines(unique(logdose),corti.nl1$response)
```

(See figure 1.6, page 15).

ELISA test example

Creating the data

The experimental data (see table 1.4, page 7) are stored in a *data-frame*. The independent variable is equal to $\log_{10}$ of the dilution. The curves are called "m" for May and "j" for June.

```
> dilution <-rep(c(
        rep(30,2), rep(90,2), rep(270,2), rep(810,2),
        rep(2430,2), rep(7290,2), rep(21869,2), rep(65609,2)), 2)
> elisa <- data.frame(logd=log10(dilution),
            OD=c(1.909, 1.956, 1.856, 1.876, 1.838, 1.841, 1.579,
                 1.584, 1.057, 1.072, 0.566, 0.561, 0.225, 0.229,
                 0.072, 0.114, 1.886, 1.880, 1.853, 1.870, 1.747,
                 1.772, 1.424, 1.406, 0.781, 0.759, 0.377, 0.376,
                 0.153, 0.138, 0.053, 0.058),
            curves=c(rep("m", 16), rep("j", 16)))
```

Plot of the observed responses versus the logarithm of the dilution

We plot the observed response values versus the logarithm of the dilution by using the function pldnls2:

```
> pldnls2(elisa,response.name="OD",X.names="logd")
```

(See figure 1.4, page 7).

Description of the model

The model defined in section 1.4.3, page 15, is described in a file called
elisa.mod1:

```
resp OD;
varind logd;
aux a1;
parresp p1,p2,p3,p4;
subroutine;
begin
a1= 1+exp(p3*(logd-p4));
OD=p1+(p2-p1)/a1;
end
```

Parameter estimation

To estimate the parameters, we use the function nls2:

```
> elisa.nl1<-nls2(elisa,"elisa.mod1",rep(c(2,0,1,0),2))
> cat( "Estimated values of the parameters:\n ")
> print( elisa.nl1$theta); cat( "\n\n")
```

(Results are given in section 1.4.3, page 15.)

Plot of the observed and fitted curves

We plot the observed and fitted response values by using the function
plfit:

```
> plfit(elisa.nl1, title="ELISA assay")
```

(See figure 1.7, page 16).

Ovocytes example

Creating the data

The *data-frame* is created from the values of the time and the volumes of
ovocytes. The two curves are called "m" for mature ovocytes and "i" for
immature ovocytes (see table 1.5, page 10.)

```
> Times <- c( rep(0.5,7),rep(1,7),rep(1.5,7),rep(2,7),
        rep(2.5,7),rep(3,7),rep(4,7),rep(5,7),
        rep(7,7),rep(10,7),rep(12,7),rep(15,7),rep(20,7),
        rep(0.5,5),rep(1,5),rep(1.5,5),rep(2,5),
        rep(2.5,5),rep(3,5),rep(3.5,5), rep(4,5), rep(4.5,5),
        rep(5,5), rep(6.5,5),rep(8,5), rep(10,5),rep(12,5),
        rep(15,5),rep(20,5))
> V <- c (0.6833, 0.6870, 0.7553, 0.6012, 0.6655, 0.7630, 0.7380,
        0.6275, 0.5291, 0.5837, 0.5425, 0.5775, 0.6640, 0.6022,
        0.6743, 0.5890, 0.5837, 0.5713, 0.6309, 0.6960, 0.6344,
        0.7290, 0.6205, 0.5912, 0.5936, 0.6932, 0.7544, 0.6849,
        0.7479, 0.6532, 0.6064, 0.6164, 0.7132, 0.7717, 0.7200,
```

```
        0.7672, 0.6870, 0.6376, 0.6884, 0.7494, 0.8070, 0.7569,
        0.7865, 0.7219, 0.6456, 0.7747, 0.8064, 0.8342, 0.8100,
        0.8265, 0.7580, 0.6782, 0.8205, 0.8554, 0.8527, 0.8424,
        0.8897, 0.8346, 0.8238, 0.8680, 0.8955, 0.8621, 0.8490,
        0.9250, 0.9000, 0.9622, 0.9477, 0.9321, 0.8809, 0.8725,
        0.9480, 0.9290, 1.0000, 0.9700, 0.9362, 0.8905, 0.8930,
        0.9820, 0.9550, 1.0000, 1.0000, 0.9784, 0.9100, 0.9242,
        1.0000, 0.9800, 1.0000, 1.0000, 1.0000, 0.9592, 0.9779,
        0.4536, 0.4690, 0.6622, 0.6210, 0.5130,
        0.4297, 0.4552, 0.5530, 0.3700, 0.4250,
        0.4876, 0.4690, 0.5530, 0.4000, 0.4500,
        0.5336, 0.4760, 0.5535, 0.4200, 0.4750,
        0.5536, 0.4830, 0.5881, 0.4600, 0.5000,
        0.5618, 0.5047, 0.6097, 0.4900, 0.5500,
        0.6065, 0.5269, 0.6319, 0.5200, 0.6000,
        0.6365, 0.5421, 0.6699, 0.5500, 0.6500,
        0.6990, 0.5814, 0.6935, 0.5700, 0.7000,
        0.7533, 0.6225, 0.7176, 0.5800, 0.7500,
        0.7823, 0.6655, 0.7422, 0.6300, 0.7900,
        0.8477, 0.7105, 0.7674, 0.6900, 0.8300,
        0.8966, 0.8265, 0.8464, 0.7500, 0.8800,
        0.9578, 0.9105, 0.9115, 0.8000, 0.9500,
        1.0000, 0.9658, 0.9404, 0.9000, 1.0000,
        1.0000, 1.0000, 0.9800, 0.9410, 1.0000)
> ovo <- data.frame(T=Times,V, curves=c(rep("m", 91), rep("i", 80)))
```

Description of the model

The model defined in section 1.4.4, page 16, is described in a file called
ovo.mod1. The known parameters are declared to be *second-level parameters*, except for the parameter called ov whose value depends on the curve.
The value of ov will be set by numerical equality constraints when calling
nls2.

```
resp V;
varind T;
parresp Pw, Ps, ov;
varint t;
valint T;
pbisresp mm, mv, mo, mme, mve, Vo, a, moe, moi, phi;
aux z1, z3, g1,g2,g3,g4,g5;
F Vw, Vs;
dF dVw, dVs;
subroutine;
begin
z1 = Vo*(1-ov)*moi/1e+3;
z3 = a*phi*mv/mm ;
g1 = -a*(mme/mve)*(moe+mo)/1e+3;
g2 = (mme/mve)*a*z1;
g3 = (mme/mve)*z3;
g4 = (mm/mv)*a*mo/1e+3;
g5 = -a*phi;
dVw = Pw*(g1 + g2/Vw + g3*Vs/Vw);
```

```
dVs = Ps*(g4 + g5*Vs/Vw);
V = Vw[T]/Vo + Vs[T]/Vo + ov;
end
```

A C-program to calculate the model

The system **nls2** includes the possibility of generating a C-program from the formal description of the model. Once compiled and loaded into the **S-PLUS** session, this program is used by the function nls2 instead of the default evaluation, that is, evaluation by syntaxical trees. Using C-programs reduces execution time and saves memory space. Here, because the model is especially complicated, we use this option.

The operating-system command analDer, provided with the system **nls2**, generates a C-routine in a file called ovo.mod1.c. Then, the file is compiled.

```
$  analDer ovo.mod1
$  Splus COMPILE ovo.mod1.c INC=$SHOME/library/nls2/include
```

The compiled file is loaded into the **S-PLUS** session by using function loadnls2:

```
> loadnls2("ovo.mod1.o")
```

Parameter estimation

To estimate the parameters, the function nls2 is called with four arguments: the name of the *data-frame*, a list that describes the model, ovo.mod1, the starting values of the parameters, and a list that gives information about the ordinary differential equation system, ovo.int.

ovo.mod1 includes the name of the description file, the values of the *second-level parameters*, and a component called eq.theta, which contains the values of the numerical equality constraints on the parameters. With these constraints, we set the parameter ov to its known value. The value NaN (NotANumber) means that no constraint is set on the corresponding parameter.

ovo.int includes the initial value of the integration variable, the number of parameters in the system and its initial values (a matrix with two rows, one for each curve, and two columns, one for each equation of the system).

```
> ovo.mod1<-  # The list that describes the model
      list(file="ovo.mod1",
      gamf=c(76.09, 1.04, 1.62, 18,1,1.596e-6,6.604e-4,0.29,0.29,1),
      eq.theta=c(NaN, NaN, 0.15, NaN, NaN, 0.065))
> ovo.int<-   # The list that describes the integration context
    list(start=0, nb.theta.odes=3,
    cond.start=matrix(c(1.35660e-6, 0, 1.49226e-06, 0), ncol=2))
> ovo.nl1<-nls2(ovo,
                ovo.mod1,
                c(0.1, 0.01, 0.15, 0.1, 0.01, 0.065),
                integ.ctx=ovo.int)
> cat( "Estimated values of the parameters:\n ")
```

```
> print( ovo.nl1$theta); cat( "\n\n")
```

(Results are given in section 1.4.4, page 16.)

Plot of the results

We plot the observed and fitted response values by using function `plfit`:

```
> plfit(ovo.nl1, title="Ovocytes example")
```

(See figure 1.8, page 17).

Isomerization example

Creating the data

The *data-frame* includes three independent variables: H for hydrogen, P for n-pentane, and I for iso-pentane. The response values are the observed rates, r (see table 1.6, page 11).

```
> isomer <- data.frame(
  H=c(205.8, 404.8, 209.7, 401.6, 224.9, 402.6, 212.7,
      406.2, 133.3, 470.9, 300.0, 301.6, 297.3, 314.0,
      305.7, 300.1, 305.4, 305.2, 300.1, 106.6, 417.2,
      251.0, 250.3, 145.1),
  P=c(90.9, 92.9, 174.9, 187.2, 92.7, 102.2, 186.9, 192.6,
      140.8, 144.2, 68.3, 214.6, 142.2, 146.7, 142.0, 143.7,
      141.1, 141.5, 83.0, 209.6, 83.9, 294.4, 148.0, 291.0),
  I=c(37.1, 36.3, 49.4, 44.9, 116.3, 128.9, 134.4, 134.9,
      87.6, 86.9, 81.7, 101.7, 10.5, 157.1, 86.0, 90.2, 87.4,
      87.0, 66.4, 33.0, 32.9, 41.5, 14.7, 50.2),
  r=c(3.541, 2.397, 6.694, 4.722, 0.593, 0.268, 2.797, 2.451,
      3.196, 2.021, 0.896, 5.084, 5.686, 1.193, 2.648, 3.303,
      3.054, 3.302, 1.271, 11.648, 2.002, 9.604, 7.754, 11.590))
```

Description of the model

The model defined in section 1.4.5, page 17, is described in a file called `isomer.mod1`:

```
resp r;
varind H,P,I;
aux a1, a2;
parresp t1,t2,t3,t4;
subroutine;
begin
a1= t1*t3*(P-I/1.632);
a2= 1+t2*H+t3*P+t4*I;
r=a1/a2;
end
```

Parameter estimation

To estimate the parameters, the function **nls2** is called. Its arguments are the name of the *data-frame*, the name of the file that describes the model and a list of information about the statistical context: the starting values of the parameters and the requested maximal number of iterations.

```
> loadnls2() # load the programs
> isomer.nl1<-nls2(isomer,"isomer.mod1",
      list(theta.start=c(10,1,1,1), max.iters=100))
> cat( "Estimated values of the parameters:\n ")
> print( isomer.nl1$theta); cat( "\n\n")
```

(Results are given in section 1.4.5, page 17.)

Plot of the fitted values versus the observed values

We plot the fitted values versus the observed values by using the function **plfit**:

```
> plfit(isomer.nl1, wanted=list(O.F=T),
        title="Isomerization example")
```

(See figure 1.9, page 18).

2

Accuracy of estimators, confidence intervals and tests

To determine the accuracy of the estimates that we made in chapter 1, we demonstrate in this chapter how to calculate confidence intervals and how to perform tests. The specific concerns of each of our particular experiments are described below. We then introduce the methodology, first describing classical asymptotic procedures and two asymptotic tests, the Wald test and the likelihood ratio test, and then presenting procedures and tests based on a resampling method, the bootstrap. As before, we conclude the chapter by applying these methods to the examples.

2.1   Examples

Let us assume that in example 1.1.1 we are interested in the maximum yield, θ_1. We have calculated one estimate of θ_1, $\widehat{\theta}_1 = 69.95$; however, if we do another experiment under the same experimental conditions, the observed values of Y and the estimates of the parameters will both be different. Thus knowing one estimate is not entirely satisfactory, and we need to quantify its accuracy.

In example 1.1.2, we calculated an estimate of the calibration curve. Suppose that we now want to estimate the dose of hormone D contained in a preparation that has the expected response $\mu = 2000$ c.p.m. To do this we must use equation (1.6), replacing the parameters by their estimates. We find $\widehat{X} = -1.1033$ and $\widehat{D} = \exp \widehat{X} \log 10 = 0.3318$ ng/.1 ml, but we now need to calculate how much confidence we can place on this estimate.

Let us consider example 1.1.3, in which we estimated two ELISA response curves. Our concern in this experiment was to estimate the relative potency of the two different serums. In order to do this, however, we must verify whether the condition of *parallelism*, as expressed in formula (1.7), is true. If it does exist for all values of x, then the parameters verify that $\theta_1^{\text{May}} = \theta_1^{\text{June}}$, $\theta_2^{\text{May}} = \theta_2^{\text{June}}$, $\theta_3^{\text{May}} = \theta_3^{\text{June}}$. In this case, $\beta = \theta_4^{\text{May}} - \theta_4^{\text{June}}$ is the horizontal distance between the two curves at the inflection point. To determine parallelism we will first test if these relations between the parameters are true, or, more exactly, if they do not contradict the data. If the test does not reject the hypothesis of parallelism, we will be able to estimate β and to test if β is significantly different from zero.

In example 1.1.4 we were interested in comparing the parameters P_w and P_s. Because water permeability in cells is higher than propane-diol permeability, water flows out of the cells more rapidly than the propane-diol flows in, resulting in high cell shrinkage. Thus we are interested in the speed of intracellular propane-diol penetration; the cryoprotectant must permeate the ovocytes in a short time. To this end, we will compare the values of V_s at times $T_1 = 1\text{mn}$, $T_2 = 5\text{mn}$, $\ldots$.

Using example 1.1.5, we will compare different methods for calculating confidence intervals for the parameters.

In sum, what we want to do in all of the above examples is determine if we estimated the function of the parameters denoted $\lambda(\theta)$ accurately. In the first example, $\lambda = \theta_1$, and in the second example, $\lambda = \exp(X \log 10)$, where X is defined by

$$ X = \frac{1}{\theta_4} \left\{ \log \left[\exp \frac{1}{\theta_5} \log \frac{\theta_2 - \theta_1}{\mu - \theta_1} - 1 \right] - \theta_3 \right\}. $$

In the third example, if the hypothesis of parallelism is not rejected, we are interested in $\lambda = \beta$ or $\lambda = \exp -\beta$. In the fourth example we are interested in the pairs (P_w, P_s) for each curve, and in $\lambda = V_s(T_1)$ and $\lambda = V_s(T_2)$.

2.2   Problem formulation

The nonlinear regression model is defined in section 1.2. Let $\widehat{\theta}$ be the least squares estimator of θ when we have homogeneous variances, and let it be the weighted least squares estimator of θ when we have heterogeneous[1] variances (see section 1.3).

Let λ be a function of the parameters and $\widehat{\lambda}$ an estimator of λ: $\widehat{\lambda} = \lambda(\widehat{\theta})$. In this chapter, we will describe how to calculate a confidence interval for λ, and how to do a test.

---

[1] See chapter 3 for a complete treatment when the variance of errors is not constant.

The function λ must verify some *regularity* assumptions. It must be a continuous function of θ with continuous partial derivatives with respect to θ.

2.3   Solutions

2.3.1   Classical asymptotic results

$\widehat{\theta}$ is a function of the Y_{ij}, and when the number of observations tends to infinity, its distribution is known: $\widehat{\theta} - \theta$ tends to 0, and the limiting distribution of $V_{\widehat{\theta}}^{-1/2}(\widehat{\theta} - \theta)$ is a standard p-dimensional normal (Gaussian) distribution $\mathcal{N}(0, I_p)$ with expectation 0 and variance I_p, where I_p is the $p \times p$ identity matrix. $V_{\widehat{\theta}}$ is the estimated asymptotic covariance matrix of $\widehat{\theta}$. Thus, for sufficiently large n, the distribution of $\widehat{\theta}$ may be approximated by the normal distribution $\mathcal{N}(\theta, V_{\widehat{\theta}})$.

We need a result for $\widehat{\lambda} = \lambda(\widehat{\theta})$. The limiting distribution (when n tends to infinity) of $\widehat{T} = (\widehat{\lambda} - \lambda)/\widehat{S}$ will be a centered normal distribution, $\mathcal{N}(0, 1)$, where $\widehat{S}$ is an estimate of the standard error of $\widehat{\lambda}$.

Notations and formulae: f_i is for $f(x_i, \theta)$. The p-vector of derivatives of f with respect to θ calculated in x_i is denoted by $\dfrac{\partial f_i}{\partial \theta}$. The components of $\dfrac{\partial f_i}{\partial \theta}$ are $\dfrac{\partial f}{\partial \theta_a}(x_i, \theta)$, $a = 1, \ldots p$.

Let Γ_θ be the $p \times p$ matrix defined as follows:

$$\Gamma_\theta = \frac{1}{\sigma^2}\frac{1}{n}\sum_{i=1}^{k} n_i \frac{\partial f_i}{\partial \theta}\left(\frac{\partial f_i}{\partial \theta}\right)^T,$$

where the exponent T means that the vector is transposed. The elements (a, b) of Γ_θ are

$$\Gamma_{\theta, ab} = \frac{1}{\sigma^2}\frac{1}{n}\sum_{i=1}^{k} n_i \frac{\partial f_i}{\partial \theta_a}\frac{\partial f_i}{\partial \theta_b}.$$

Let Δ_θ be the $p \times p$ matrix

$$\frac{1}{n}\sum_{i=1}^{k} \frac{n_i}{\sigma_i^2}\frac{\partial f_i}{\partial \theta}\left(\frac{\partial f_i}{\partial \theta}\right)^T.$$

Let $\widehat{f}_i = f(x_i, \widehat{\theta})$, and $\dfrac{\partial \widehat{f}_i}{\partial \theta}$ be the vector with components $\dfrac{\partial f}{\partial \theta_a}(x_i, \widehat{\theta})$, and let $\Gamma_{\widehat{\theta}}$ and $\Delta_{\widehat{\theta}}$ be the matrices Γ_θ and Δ_θ, where the unknown parameters are replaced by their estimators:

$$\Gamma_{\widehat{\theta}} = \frac{1}{\widehat{\sigma}^2}\frac{1}{n}\sum_{i=1}^{k} n_i \frac{\partial \widehat{f}_i}{\partial \theta}\left(\frac{\partial \widehat{f}_i}{\partial \theta}\right)^T, \quad \Delta_{\widehat{\theta}} = \frac{1}{n}\sum_{i=1}^{k} \frac{n_i}{\widehat{\sigma}_i^2}\frac{\partial \widehat{f}_i}{\partial \theta}\left(\frac{\partial \widehat{f}_i}{\partial \theta}\right)^T,$$

and $\widehat{\sigma}^2 = C(\widehat{\theta})/n$, $\widehat{\sigma}_i^2 = s_i^2$.

$V_{\widehat{\theta}}$ is the estimate of V_θ, the $p \times p$ asymptotic covariance matrix of $\widehat{\theta}$:

in the case $\mathrm{Var}(\varepsilon_{ij}) = \sigma^2$, $V_{\widehat{\theta}} = \dfrac{1}{n}\Gamma_{\widehat{\theta}}^{-1}$,

in the case $\mathrm{Var}(\varepsilon_{ij}) = \sigma_i^2$, $V_{\widehat{\theta}} = \dfrac{1}{n}\Delta_{\widehat{\theta}}^{-1}$.

Because $\widehat{\theta} - \theta$ is small, we get the limiting distribution of $\lambda(\widehat{\theta})$ by approximating $\lambda(\widehat{\theta}) - \lambda(\theta)$ by a linear function of $\widehat{\theta} - \theta$:

$$\left(\frac{\partial\lambda}{\partial\theta}\right)^T (\widehat{\theta} - \theta) = \sum_{a=1}^{p} \frac{\partial\lambda}{\partial\theta_a}(\widehat{\theta}_a - \theta_a).$$

Thus $(\widehat{\lambda} - \lambda)/S_{\widehat{\theta}}$ is distributed, when n tends to infinity, as a $\mathcal{N}(0,1)$ (a Gaussian centered variate with variance 1), where

$$S_\theta^2 = \left(\frac{\partial\lambda}{\partial\theta}\right)^T V_\theta \frac{\partial\lambda}{\partial\theta} = \sum_{a=1}^{p}\sum_{b=1}^{p} \frac{\partial\lambda}{\partial\theta_a}\frac{\partial\lambda}{\partial\theta_b} V_{\theta,ab} \qquad (2.1)$$

and $\widehat{S} = S_{\widehat{\theta}}$ is the asymptotic estimate of the standard error.

2.3.2 *Asymptotic confidence intervals for* λ

If the distribution of $\widehat{T}$ were known, say $F(u) = \mathrm{Pr}(\widehat{T} \leq u)$, we would calculate the $\alpha/2$- and $1 - \alpha/2$-percentiles[2] of $\widehat{T}$, say u_α, $u_{1-\alpha/2}$. The interval

$$\widehat{I} = \left[\widehat{\lambda} - u_{1-\alpha/2}\widehat{S} \ ; \ \widehat{\lambda} - u_{\alpha/2}\widehat{S}\right]$$

would be a confidence interval for λ, with level $1 - \alpha$. In this case, the coverage probability of $\widehat{I}$, the probability that $\widehat{I}$ covers λ, would be $1 - \alpha$.

However, as we have seen in the preceding paragraph, we can only approximate, when n is sufficiently large, the distribution of $\widehat{T}$. Thus we use this approximation to calculate confidence intervals with coverage probability close to $1 - \alpha$.

Let $\mathcal{N}$ be a variate distributed as a $\mathcal{N}(0,1)$. Let ν_α be the α-percentile of $\mathcal{N}$. From the result of section 2.3.1, we can deduce a confidence interval for λ:

$$\widehat{I}_{\mathcal{N}} = \left[\widehat{\lambda} - \nu_{1-\alpha/2}\widehat{S}; \widehat{\lambda} + \nu_{1-\alpha/2}\widehat{S}\right]. \qquad (2.2)$$

This interval is symmetric around $\widehat{\lambda}$ for $\nu_\alpha = -\nu_{1-\alpha}$.

The probability that $\widehat{T}$ is less than ν_α tends to α when n tends to infinity; the probability for λ to lie in $\widehat{I}_{\mathcal{N}}$ tends to $1 - \alpha$ when n tends to infinity. We say that $\widehat{I}_{\mathcal{N}}$ has asymptotic level $1 - \alpha$.

---

[2]The α-percentile of a variate with distribution function F is the value of u, say u_α, such that $F(u_\alpha) = \alpha$ and $0 < \alpha < 1$.

Remarks.

1. By analogy to the Gaussian linear regression case, in the nonlinear regression model with homogeneous variance, we define an alternative confidence interval for λ. In formula (2.2) we replace ν_α by $\sqrt{n/(n-p)}t_\alpha$, where t_α is the α-percentile of a Student variate with $n-p$ degrees of freedom:

$$\widehat{I}_T = \left[\widehat{\lambda} - \sqrt{\frac{n}{n-p}}t_{1-\alpha/2}\widehat{S}; \widehat{\lambda} + \sqrt{\frac{n}{n-p}}t_{1-\alpha/2}\widehat{S}\right]. \qquad (2.3)$$

$\widehat{I}_T$ has the same asymptotic level as $\widehat{I}_N$, but $\widehat{I}_T$ is wider than $\widehat{I}_N$ and its coverage probability will be greater. Some studies [HJM89] have shown that $\widehat{I}_T$ has a coverage probability closer to $1-\alpha$ than $\widehat{I}_N$.

2. The intervals $\widehat{I}_N$ and $\widehat{I}_T$ are symmetric around $\widehat{\lambda}$. In some applications, a part of the symmetric confidence interval might not coincide with the set of variations of the parameter λ. For example, consider $\lambda = \exp\theta_3$ in the pasture regrowth example. If the estimate of the standard error of $\widehat{\lambda}$, say $\widehat{S}$, is bigger than $\widehat{\lambda}/\nu_{1-\alpha/2}$, then the lower bound of $\widehat{I}_N$ is negative even though λ is strictly positive. In that case, it is easy to see that it is more appropriate to calculate a confidence interval for θ_3 and then to transform this interval taking the exponential of its limits to find a confidence interval for λ. More generally, let $\widehat{S}_3$ be the estimate of the standard error of $\widehat{\theta}_3$, and let g be a strictly increasing function of θ_3. If θ_3 lies in

$$\left[\widehat{\theta}_3 - \nu_{1-\alpha/2}\widehat{S}_3; \widehat{\theta}_3 + \nu_{1-\alpha/2}\widehat{S}_3\right],$$

then $\lambda = g(\theta_3)$ lies in

$$\left[g(\widehat{\theta}_3 - \nu_{1-\alpha/2}\widehat{S}_3); g(\widehat{\theta}_3 + \nu_{1-\alpha/2}\widehat{S}_3)\right].$$

2.3.3 *Asymptotic tests of "$\lambda = \lambda_0$" against "$\lambda \neq \lambda_0$"*

Let λ_0 be a fixed value of λ and let the hypothesis of interest be "H: $\lambda = \lambda_0$" against "A: $\lambda \neq \lambda_0$".

Wald test. When H is true the limiting distribution of $(\widehat{\lambda} - \lambda_0)/\widehat{S}$ is a $\mathcal{N}(0,1)$. Thus, the limiting distribution of the test statistic

$$\mathcal{S}_W = \frac{(\widehat{\lambda} - \lambda_0)^2}{\widehat{S}^2}$$

is a χ^2 with 1 degree of freedom. The hypothesis H will be rejected for large values of $\mathcal{S}_W$, say $\mathcal{S}_W > C$, where C is chosen such that $\Pr(Z_1 \leq C) = 1-\alpha$, Z_1 being distributed as a χ^2 with 1 degree of freedom.

This test is called the Wald test. When H is true, the probability for $\mathcal{S}_W$ to be greater than C (in other words, the probability that the hypothesis H is rejected when it should be accepted) tends to α when n tends to infinity. We say that this test has asymptotic error of the first kind equal to α. Assume now that H is false. Then the power of the test defined as the probability for $\mathcal{S}_W$ to be greater than C (in other words, the probability to reject the hypothesis H when H is false) tends to 1 when n tends to infinity. We say that this test is consistent.

Remark. As in section 2.3.2, homogeneous variances can be considered separately; the hypothesis H will be rejected if

$$\frac{n-p}{n}\mathcal{S}_W > C,$$

where C is chosen such that $\Pr(F_{1,n-p} \leq C) = 1 - \alpha$, $F_{1,n-p}$ being distributed as a Fisher variable with 1 and $n - p$ degrees of freedom.

Likelihood ratio type test. Another idea is to estimate the parameters under the constraint $\lambda = \lambda_0$, say $\widehat{\theta}_H$, then to estimate them without the constraint, say $\widehat{\theta}_A$, and then to compare the estimation criterions (1.10) $C(\widehat{\theta}_H)$ and $C(\widehat{\theta}_A)$ in the case of homogeneous variances. If H is true, the difference between $C(\widehat{\theta}_H)$ and $C(\widehat{\theta}_A)$ will be small. Let

$$\mathcal{S}_L = n \log C(\widehat{\theta}_H) - n \log C(\widehat{\theta}_A)$$

be the test statistic. When n tends to infinity, it can be shown that the limiting distribution of $\mathcal{S}_L$ is a χ^2 with 1 degree of freedom. The hypothesis H will be rejected when $\mathcal{S}_L > C$, where C is chosen such that $\Pr(Z_1 \leq C) = 1 - \alpha$.

This test based on $\mathcal{S}_L$ is called a likelihood ratio type test. This test has the same asymptotic properties as the Wald test. Although the Wald test is easier to calculate, some theoretical arguments favor the likelihood ratio test.

2.3.4 *Asymptotic tests of "$\Lambda\theta = L_0$" against "$\Lambda\theta \neq L_0$"*

Let us return to example 1.1.5 and assume that we want to test if the parameters θ_2, θ_3 and θ_4 are identical. The hypothesis of interest is "H : $\theta_2 = \theta_3 = \theta_4$" against the alternative that at least two of these parameters are different. H can be written as "$\Lambda\theta = 0$," where Λ is the following 2×4 matrix:

$$\Lambda = \begin{pmatrix} 0 & 1 & -1 & 0 \\ 0 & 1 & 0 & -1 \end{pmatrix}. \tag{2.4}$$

The problems defined above can be solved by returning to the general case, with θ of dimension p. We aim to test the hypothesis "H: $\Lambda\theta = L_0$" against "A: $\Lambda\theta \neq L_0$," where Λ is a $q \times p$ matrix of rank q, $q < p$, and L_0 is a vector of dimension q. The model defined by the hypothesis H is a model nested in the more general one defined by the hypothesis A.

The Wald test. When H is true, the limiting distribution of

$$\mathcal{S}' = (\Lambda V_{\widehat{\theta}}\Lambda^T)^{-1/2}(\Lambda\widehat{\theta} - L_0)$$

is a q-dimensional Gaussian variable with mean 0 and covariance matrix equal to the $q \times q$ identity matrix. In these cases, the limiting distribution of the test statistic $\mathcal{S}_W = \sum_{a=1}^q \mathcal{S}_a'^2$ is a χ^2 with q degrees of freedom. The Wald test is defined by the rejection of H when $\mathcal{S}_W > C$, where $\Pr(Z_q \le C) = 1 - \alpha$, Z_q being distributed as a χ^2 with q degrees of freedom.

Remark. In the case of homogeneous variances, the test is defined by the rejection of H when

$$\frac{n-p}{n}\frac{\mathcal{S}_W}{q} > C, \tag{2.5}$$

where C is chosen such that $\Pr(F_{q,n-p} \le C) = 1 - \alpha$, $F_{q,n-p}$ being distributed as a Fisher with q and $n - p$ degrees of freedom.

The likelihood ratio test. Let $\widehat{\theta}_H$ be the estimation of θ under the constraint "$\Lambda\theta = L_0$"; then, in the case of homogeneous variances, the limiting distribution of the test statistic $\mathcal{S}_L = n \log C(\widehat{\theta}_H) - n \log C(\widehat{\theta}_A)$ is a χ^2 with q degrees of freedom. This result provides the likelihood ratio type test.

Curve comparison. Let us return to example 1.1.3, where we needed to compare two curves. The hypothesis of interest is "H: $\theta_1^{\text{May}} = \theta_1^{\text{June}}$, $\theta_2^{\text{May}} = \theta_2^{\text{June}}$, $\theta_3^{\text{May}} = \theta_3^{\text{June}}$" against the alternative that at least one of these equalities is false. We create a data set by joining the data observed in May and June. We define the vector of parameters by joining θ^{May} and θ^{June}: let

$$\theta = \begin{pmatrix} \theta^{\text{May}} \\ \theta^{\text{June}} \end{pmatrix}$$

be the $2p$-vector of parameters for the two curves. Then the hypothesis H can be written as above, "$\Lambda\theta = 0$", where Λ is the following $3 \times 2p$ matrix:

$$\Lambda = \begin{pmatrix} 1 & 0 & 0 & 0 & -1 & 0 & 0 & 0 \\ 0 & 1 & 0 & 0 & 0 & -1 & 0 & 0 \\ 0 & 0 & 1 & 0 & 0 & 0 & -1 & 0 \end{pmatrix}. \tag{2.6}$$

We define a test using as before the statistic $\mathcal{S}_W$ or $\mathcal{S}_L$.

2.3.5 Bootstrap estimations

Resampling methods like the jackknife and the bootstrap are especially useful for estimating the accuracy of an estimator. We observe $Y_1, Y_2, \ldots Y_n$;

we choose a parametric nonlinear regression model with parameters θ, and we find an estimation procedure to estimate a function of θ, say $\lambda(\theta)$. We get $\widehat{\lambda} = \lambda(\widehat{\theta})$, but we are interested in calculating the accuracy of $\widehat{\lambda}$ or, more generally, in knowing its distribution (or some characteristics of it). If we were able to repeat the experiment under exactly the same conditions, we would observe $Y_1^1, Y_2^1, \ldots Y_n^1$, and in the same way as for $\widehat{\lambda}$ we would calculate $\widehat{\lambda}^1$. We could repeat it again, and calculate $\widehat{\lambda}^2$ with $Y_1^2, Y_2^2, \ldots Y_n^2$. $\widehat{\lambda}^1, \widehat{\lambda}^2, \ldots$ would be a sample of random variables distributed as $\widehat{\lambda}$. This sample would approximate the distribution of $\widehat{\lambda}$.

In short, resampling methods are a way to mimic the repetition of the experiment. Here we will focus only on the bootstrap method in the case of homogeneous variances.

Bootstrap estimations are based on estimates $\widehat{\lambda}^\star = \lambda(\widehat{\theta}^\star)$ calculated from artificial bootstrap samples $(x_i, Y_{ij}^\star)$, $j = 1, \ldots, n_i$, $i = 1, \ldots k$, where

$$Y_{ij}^\star = f(x_i, \widehat{\theta}) + \varepsilon_{ij}^\star.$$

The errors $\varepsilon_{ij}^\star$ are simulated in the following way: let $\widehat{\varepsilon}_{ij} = Y_{ij} - f(x_i, \widehat{\theta})$ be the residuals, and let $\widetilde{\varepsilon}_{ij} = \widehat{\varepsilon}_{ij} - \widehat{\varepsilon}_\bullet$, be the centered residuals, $\widehat{\varepsilon}_\bullet$ being the sample mean, $\widehat{\varepsilon}_\bullet = \sum_{i,j} \widehat{\varepsilon}_{ij}/n$. The set of $\varepsilon_{ij}^\star$, for $j = 1, \ldots n_i$, and $i = 1, \ldots k$ is a random sample from the empirical distribution function based on the $\widetilde{\varepsilon}_{ij}$ (n $\widetilde{\varepsilon}_{ij}$ are drawn with replacement, each with probability $1/n$). There are n^n such different samples.

$\widehat{\theta}^\star$ will be the value of θ that minimizes

$$C^\star(\theta) = \sum_{i=1}^{k} \sum_{j=1}^{n_i} (Y_{ij}^\star - f(x_i, \theta))^2.$$

The bootstrap estimate of λ is $\widehat{\lambda}^\star = \lambda(\widehat{\theta}^\star)$.

Let B be the number of bootstrap simulations. $(\widehat{\lambda}^{\star,b} = \lambda(\widehat{\theta}^{\star,b}), b = 1, \ldots, B)$ is a B-sample of bootstrap estimates of λ. The choice of B will be discussed later, at the end of this section. The important result is that the distribution of $\widehat{\lambda}^\star$, estimated by the empirical[3] distribution function of the $(\widehat{\lambda}^{\star,b}, b = 1, \ldots B)$, approximates the distribution of $\widehat{\lambda}$. Let

$$\widehat{T}^\star = \frac{\widehat{\lambda}^\star - \widehat{\lambda}}{S_{\widehat{\theta}^\star}}.$$

Roughly speaking, the difference between the distribution functions of $\widehat{T}$ and $\widehat{T}^\star$ tends to 0 when the number of observations n is large; thus, we can use the quantiles of $\widehat{T}^\star$ instead of those of $\widehat{T}$ to construct confidence intervals or tests.

---

[3] Obviously, B must be large enough so that the empirical distribution function is a good approximation of the distribution of $\widehat{\lambda}^\star$. If $B = n^n$, and if we draw all of the possible samples, we get the exact distribution of $\widehat{\lambda}^\star$.

Bootstrap confidence interval for λ

Let $(\widehat{T}^{*,b}, b = 1, \ldots B)$ be a B-sample of $\widehat{T}^*$; $\widehat{T}^*$ is calculated in the same way as $\widehat{T}$, replacing Y_{ij} by Y_{ij}^*. Let b_α be the α-percentile of the $\widehat{T}^{*,b}$ (the way of calculating b_α is detailed in section 2.4.1). It can be shown that $\Pr(\widehat{T} \le b_\alpha)$ tends to α when n tends to infinity. This gives a bootstrap confidence interval for λ:

$$\widehat{I}_B = \left[\widehat{\lambda} - b_{1-\alpha/2}\widehat{S}; \widehat{\lambda} - b_{\alpha/2}\widehat{S}\right]. \tag{2.7}$$

For large n and B the coverage probability of $\widehat{I}_B$ is close to $1 - \alpha$.

Bootstrap estimation of the accuracy of $\widehat{\lambda}$

The variance and even the bias of $\widehat{\lambda}$ may be infinite or undefined. Nevertheless, their estimates (using the asymptotic results of section 2.3.2 or the bootstrap) measure the localization and the dispersion of the distribution of $\widehat{\lambda}$.

Variance. The bootstrap estimation of the variance is calculated by using the empirical variance of the B-sample $(\widehat{\lambda}^{*,b}, b = 1, \ldots, B)$:

$$\widehat{S}^{*2} = \sum_{b=1}^{B} \frac{1}{B-1} \left(\widehat{\lambda}^{*,b} - \widehat{\lambda}^{*,\bullet}\right)^2, \tag{2.8}$$

where $\widehat{\lambda}^{*,\bullet}$ is the sample mean $\widehat{\lambda}^{*,\bullet} = \sum_{b=1}^{B} \widehat{\lambda}^{*,b}/B$.

Bias. As we noted in section 2.3.2, the expectation of $\widehat{\lambda}$, $\mathrm{E}(\widehat{\lambda})$, is close to λ when we have large values of n. In other words, the bias of $\widehat{\lambda}$, BIAS $= \mathrm{E}(\widehat{\lambda}) - \lambda$, is close to 0. We can use the bootstrap sample to estimate this bias:

$$\widehat{\mathrm{BIAS}}^* = \widehat{\lambda}^{*,\bullet} - \widehat{\lambda} \tag{2.9}$$

Mean square error. We can estimate the mean square error (MSE) in a similar way: MSE $= \mathrm{E}(\widehat{\lambda} - \lambda)^2 = S^2 + \mathrm{BIAS}^2$, where $S^2 = \mathrm{E}(\widehat{\lambda} - \mathrm{E}(\widehat{\lambda}))^2$ is the variance of $\widehat{\lambda}$; it is estimated by

$$\widehat{\mathrm{MSE}}^* = \widehat{S}^{*2} + \widehat{\mathrm{BIAS}}^{*2}.$$

Median. Because it is always defined, the median error, the median of $\widehat{\lambda} - \lambda$, is of special interest. Its bootstrap estimate, $\widehat{\mathrm{MED}}^*$, is the median of the B values $|\widehat{\lambda}^{*,b} - \widehat{\lambda}|$.

Remarks

1. We have seen that the number of different bootstrap samples equals n^n. Obviously, we never choose for B such a value that rapidly becomes unusable ($8^8 = 16\,777\,216$!). In practice, however, a moderate number usually suffices: B around 50 in order to estimate the

accuracy characteristic, and B around 200 in order to calculate a confidence interval, for example.

2. Other resampling methods, like the jackknife, are available specially to estimate the accuracy characteristics (see [Wu86] and [Bun90] for details). These methods are less reliable than the bootstrap, however.

3. We do not discuss here the bootstrap method in the case of heterogeneous variances because the way of simulating the bootstrap samples (x_i, Y_{ij}^*), $j = 1, \ldots, n_i$, $i = 1, \ldots k$ is still under investigation and its reliability is not yet proven.

2.4   Applications

2.4.1   Pasture regrowth: calculation of a confidence interval for the maximum yield

Model. The regression function is

$$f(x, \theta) = \theta_1 - \theta_2 \exp\left(-\exp(\theta_3 + \theta_4 \log x)\right),$$

and the variances are homogeneous: $\mathrm{Var}(\varepsilon_i) = \sigma^2$.

Results.

Parameters	Estimated values	Asymptotic covariance matrix			
θ_1	69.95	3.09			
θ_2	61.68	3.87	6.66		
θ_3	-9.209	0.76	1.25	0.37	
θ_4	2.378	-0.22	-0.35	-0.09	0.027
σ^2	0.9306				

The parameter of interest is $\lambda(\theta) = \theta_1$.

Calculation of confidence intervals with asymptotic level 95%, using results of section 2.3.2.

$\widehat{\lambda}$	$\widehat{S}$	$\nu_{0.975}$	$\widehat{I}_\mathcal{N}$	$t_{0.975}$ (5 d.f.)	$\widehat{I}_T$
69.95	1.76	1.96	[66.5 , 73.4]	2.57	[63.9 , 76.0]

d.f. is for degree of freedom

Calculation of confidence intervals with asymptotic level 95%, using bootstrap. Table 2.1 gives the estimated values of f, the $\widehat{f}_i$, and the centered residuals $\widetilde{\varepsilon}_i$. For two bootstrap simulations, the table gives the bootstrap errors ε_i^*, the bootstrap observations Y_i^*, the bootstrap estimate of θ_1 and the corresponding asymptotic variance $S_{\widehat{\theta}^*}$.

B, the number of bootstrap simulations, equals 199. The histogram of the $\widehat{T}^{*,b}$, $b = 1, \ldots B$, is shown in figure 2.1.

TABLE 2.1. Results for two bootstrap simulations

$\widehat{f}_i$	$\widetilde{\varepsilon}_i$	$\varepsilon_i^{\star,1}$	$Y_i^{\star,1}$	$\varepsilon_i^{\star,2}$	$Y_i^{\star,2}$
9.411	-0.481	-0.481	8.93	-0.734	8.677
11.47	-0.669	-0.067	11.4	0.025	11.49
16.30	2.284	-0.734	15.57	-0.5	15.80
23.17	-0.843	0.025	23.20	-0.5	22.67
40.08	-0.734	-0.669	39.41	-0.481	39.60
56.18	-0.067	2.284	58.46	-0.669	55.51
60.74	0.986	0.025	60.77	-0.843	59.9
64.59	0.025	-0.734	63.86	-0.734	63.86
67.58	-0.5	2.284	69.86	2.284	69.86
$\widehat{\theta}_1 = 69.95$		$\widehat{\theta}_1^{\star,1} = 71.65$		$\widehat{\theta}_1^{\star,2} = 74.92$	
$\widehat{S} = 1.76$		$S_{\widehat{\theta}^{\star,1}} = 1.91$		$S_{\widehat{\theta}^{\star,2}} = 1.59$	

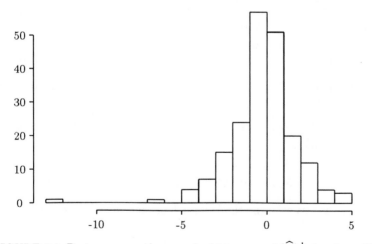

FIGURE 2.1. Pasture regrowth example: histogram of $(\widehat{T}^{\star,b}, b = 1, \ldots B)$

Calculation of the percentiles of $(\widehat{T}^{*,b}, b = 1, \ldots B)$. We calculate the 0.025 and 0.975 percentiles of the $\widehat{T}^{*,b}$ as follows: let $\widehat{T}^{*,(b)}$ be the ordered values of $\widehat{T}^{*,b}$ so that $\widehat{T}^{*,(1)} \leq \widehat{T}^{*,(2)} \leq \ldots \leq \widehat{T}^{*,(199)}$; b_α is $\widehat{T}^{*,(q_\alpha)}$, where q_α is the smallest integer such that q_α/B is greater than or equal to α. When $B = 199$, we find $b_{0.025} = \widehat{T}^{*,(5)}$ and $b_{0.975} = \widehat{T}^{*,(195)}$.

$\widehat{\lambda}$	$\widehat{S}$	$b_{0.025}$	$b_{0.975}$	$\widehat{I}_B$
69.95	1.76	-4.19	3.73	[63.4 , 77.3]

Bootstrap estimate of the accuracy characteristics.

$\widehat{\mathrm{BIAS}}^*$	$\widehat{S}*$	$\widehat{\mathrm{MSE}}^*$	$\widehat{\mathrm{MED}}^*$
0.378 (0.5% of $\widehat{\theta}^1$)	2.30	5.42	69.78

2.4.2 *Cortisol assay: estimation of the accuracy of the estimated dose* $\widehat{D}$

Model. The regression function is

$$f(x, \theta) = \theta_1 + \frac{\theta_2 - \theta_1}{(1 + \exp(\theta_3 + \theta_4 x))^{\theta_5}},$$

and the variances are heterogeneous: $\mathrm{Var}(\varepsilon_i) = \sigma_i^2$.

Results.

	Estimates	Asymptotic covariance matrix				
θ_1	133.30	0.727				
θ_2	2759.8	0.264	801.			
θ_3	3.0057	-0.0137	-2.34	0.0338		
θ_4	3.1497	-0.00723	-2.33	0.0241	0.01845	
θ_5	0.64309	0.00341	0.568	-0.00714	-0.00516	0.00152

The parameter of interest is $D = \lambda(\theta) = 10^{f^{-1}(\mu,\theta)}$; see equation (1.6), with $\mu = 2000$.

Calculation of confidence intervals with asymptotic level 95%, using results of section 2.3.2.

$\widehat{D}$	$\widehat{S}$	$\nu_{0.975}$	$\widehat{I}_N$
0.0856	0.00175	1.96	[0.0822 , 0.0891]

We will discuss other methods for calculating the accuracy of $\widehat{D}$ in chapter 5.

2.4.3 *ELISA test: comparison of curves*

We want to test the *parallelism* of the response curves in order to be able to estimate the difference $\beta = \theta_4^{\mathrm{May}} - \theta_4^{\mathrm{June}}$. We can do this by testing the hypothesis H: "$\Lambda\theta = 0$" against the alternative A: "$\Lambda\theta \neq 0$," where Λ is the matrix defined by equation (2.6),

Model. The regression function is

$$f(x,\theta) = \theta_1 + \frac{\theta_2 - \theta_1}{1 + \exp\theta_3\,(x - \theta_4)},$$

and the variances are homogeneous: $\mathrm{Var}(\varepsilon_i) = \sigma^2$.

Results.

Estimated values

θ_1^{May}	θ_2^{May}	θ_3^{May}	θ_4^{May}	θ_1^{June}	θ_2^{June}	θ_3^{June}	θ_4^{June}
0.04279	1.936	2.568	3.467	0.0581	1.909	2.836	3.251

Asymptotic covariance matrix ($\times 10^4$)

4.51							
-1.01	1.81						
15.2	-8.22	95.7					
-2.09	-0.502	-4.10	2.43				
0	0	0	0	2.51			
0	0	0	0	-0.647	1.92		
0	0	0	0	10.5	-8.74	106	
0	0	0	0	-1.05	-0.727	-0.988	1.83

$\widehat{\sigma}^2 = 0.0005602.$

Wald test of "H: $\Lambda\theta = 0$" against "A: $\Lambda\theta \neq 0$". See section 2.3.3.

The value of the Wald test is $\mathcal{S}_W = 4.53$. This number must be compared to 7.8, which is the 0.95 quantile of a χ^2 with 3 degrees of freedom. The hypothesis of *parallelism* is not rejected.

Because the variance of errors is constant, we can compare

$$\frac{n-p}{n}\frac{\mathcal{S}_W}{q} = \frac{32-8}{32}\frac{7.8}{3} = 1.133$$

to 3, which is the 0.95 quantile of a Fisher with 3 and 24 degrees of freedom.

Likelihood ratio tests. See section 2.3.3.

The two first columns of table 2.2 show the estimated parameters under A, under the constraints defined by H, and also the corresponding values of the sum of squares C.

The test statistic $\mathcal{S}_L = 32 * (\log(0.0206) - \log(0.0179))$ equals 4.5. Thus the hypothesis H is not rejected.

The estimated value of $\beta = \theta_4^{\mathrm{May}} - \theta_4^{\mathrm{June}}$ is $\widehat{\beta} = 0.223$. We can carry out a likelihood ratio type test by comparing $C(\widehat{\theta}_H)$ with $C(\widehat{\theta}_{``\beta=0"})$, which is the sum of squares when the parameters are estimated under the constraint "$\beta = 0$"; see the third column of table 2.2. $\mathcal{S}_L = 32 * (\log(0.183) - \log(0.0206)) = 69.9$. This number must be compared to 3.8, the 0.95 quantile of a χ^2 with 1 degree of freedom. The hypothesis "$\beta = 0$" is rejected.

TABLE 2.2. ELISA test example: estimated parameters under hypothesis A: the *parallelism* is not verified, hypothesis H: the curves are *parallel*, and under hypothesis "$\beta = 0$"

	Under A	Under H	Under $\beta = 0$
θ_1^{May}	0.0428	0.0501	0.0504
θ_2^{May}	1.936	1.924	1.926
θ_3^{May}	2.568	2.688	2.635
θ_4^{May}	3.467	3.470	3.356
θ_1^{June}	0.058	0.0501	0.0504
θ_2^{June}	1.909	1.924	1.926
θ_3^{June}	2.836	2.688	2.635
θ_4^{June}	3.251	3.247	3.356
$C(\theta)$	0.0179	0.0206	0.183

Conclusion of the test. In this experiment, we conclude that the *potency* ρ of the serum taken in June *relative* to the serum taken in May is estimated by $\widehat{\rho} = 10^{-\widehat{\beta}} = 0.59$, and that ρ is significantly different from 1.

Calculation of a confidence interval for ρ. The parameter of interest is $\rho = \lambda(\theta) = 10^{\theta_4^{June} - \theta_4^{May}}$.

Calculation of confidence intervals with asymptotic level 95%, using results of section 2.3.2.

$\widehat{\rho}$	$\widehat{S}$	$\nu_{0.975}$	$\widehat{I}_{\mathcal{N}}$	$t_{0.975}$ (27 d.f.)	$\widehat{I}_T$
0.59	0.0192	1.96	[0.561 , 0.636]	2.05	[0.555 , 0.642]

d.f. is for degree of freedom

$\widehat{I}_T$ is not much different from $\widehat{I}_{\mathcal{N}}$ because $n - p$ is large, $n/(n - p)$ is closed to 1, and the difference between ν_α and t_α is small.

Calculation of confidence intervals with asymptotic level 95%, using bootstrap. The number of bootstrap simulations is $B = 199$. The histogram of the $\widehat{T}^{*,b}$, $b = 1, \ldots B$, is shown in figure 2.2.
   The results are the following:

$\widehat{\rho}$	$\widehat{S}$	$b_{0.025}$	$b_{0.975}$	$\widehat{I}_B$
0.59	0.0192	-2.61	2.17	[0.557 , 0.649]

In this example, the bootstrap shows that $\widehat{I}_B$ is longer than $\widehat{I}_{\mathcal{N}}$ but is nearly equal to $\widehat{I}_T$. Moreover, the bootstrap estimation of the standard error of $\widehat{\rho}$ is $\widehat{S}^* = 0.0199$. The differences between the methods are not very important from a practical point of view.

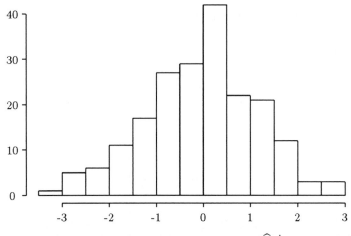

FIGURE 2.2. ELISA test example: histogram of $(\widehat{T}^{*,b}, b = 1, \ldots B)$

2.4.4   Ovocytes: calculation of confidence regions

Although this experiment tested several cryoprotectants in different experimental conditions (with or without treatment, at several temperatures) and yielded 15 curves to be estimated and compared, we present here the results for only two curves. We compute two types of confidence regions: confidence ellipsoids and likelihood contours.

Confidence ellipsoids. Let θ be the pair (P_w, P_s). When the number of observations tends to infinity, the limiting distribution of

$$\mathcal{S}'(\theta) = V_{\widehat{\theta}}^{-1/2}(\widehat{\theta} - \theta)$$

is a standard 2-dimensional normal distribution $\mathcal{N}(0, I_2)$, or the limiting distribution of

$$\mathcal{S}_W(\theta) = \mathcal{S}_1'^2 + \mathcal{S}_2'^2$$

is a χ^2 with 2 degrees of freedom. Let $r_\alpha(2)$ be the α-percentile of a χ^2 with 2 degrees of freedom, and let $\mathcal{R}_W$ be the set of θ such that $\mathcal{S}_W(\theta) \leq r_{1-\alpha}(2)$. $\mathcal{R}_W$ is an ellipse that covers θ with probability close to $1 - \alpha$.

Likelihood contours. Constructing confidence ellipses is based on the limiting distribution of $\widehat{\theta} - \theta$. Another way to calculate confidence regions for θ is to consider the limiting distribution of the statistic

$$\mathcal{S}_L(\theta) = n \log C(\theta) - n \log C(\widehat{\theta}).$$

$\mathcal{S}_L(\theta)$ is a χ^2 with 2 degrees of freedom. Let $\mathcal{R}_L$ be the set of θ such that $\mathcal{S}_L(\theta) \leq r_{1-\alpha}(2)$. $\mathcal{R}_L$ is a region of the plane that covers θ with probability close to $1 - \alpha$.

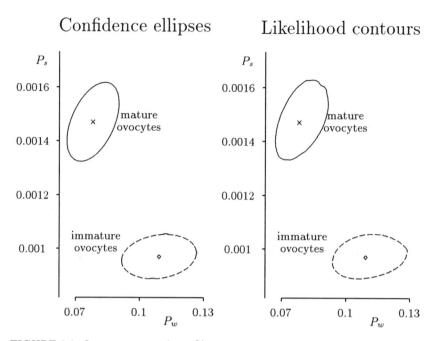

FIGURE 2.3. Ovocytes example: 95% confidence ellipses and likelihood contours for the parameters (P_w, P_s)

Figure 2.3 illustrates the confidence ellipses and the likelihood contours with level 95% for the parameters (P_w, P_s) in example 1.1.4. In this example, the likelihood contours are close to the ellipses, but we will see that this is not always the case. In fact, the discrepancy between these regions is related to the discrepancy between the distribution of $\hat{\theta} - \theta$ and its approximation by a centered Gaussian variable.

Remarks. When p, the dimension of θ, is greater than 2, the confidence ellipsoids for θ are defined by the set of θ such that $\mathcal{S}_W(\theta) \le r_{1-\alpha}(p)$, where $r_\alpha(p)$ is the α-percentile of a χ^2 with p degrees of freedom. Usually the sections of the regions are drawn in two dimensions, and they give conditional information. Consider the case $p = 3$. The sections of the confidence regions in the plane (θ_1, θ_2) are the sets of (θ_1, θ_2) such that $\mathcal{S}_W(\theta_1, \theta_2, \hat{\theta}_3) \le r_{1-\alpha}(3)$.

2.4.5 Isomerization: an awkward example

Model. The regression function is

$$f(x, \theta) = \frac{\theta_1 \theta_3 (P - I/1.632)}{1 + \theta_2 H + \theta_3 P + \theta_4 I},$$

and the variances are homogeneous: $\mathrm{Var}(\varepsilon_i) = \sigma^2$.

TABLE 2.3. Isomerization example. Estimated parameters and standard errors. Normal confidence intervals

	Estimates	Standard errors $(\widehat{S})$	95% confidence interval $\widehat{I}_N$
θ_1	35.9193	7.49	[21.20 50.60]
θ_2	0.0708583	0.163	[-0.249 0.391]
θ_3	0.0377385	0.0913	[-0.141 0.217]
θ_4	0.167166	0.379	[-0.577 0.911]
σ^2	0.13477		

Calculation of confidence intervals using the percentiles of a Gaussian distribution for each parameter. The estimated values of the parameters, of their standard errors, and of the confidence intervals calculated using formula (2.2) are given in table 2.3.

The standard errors are so large for parameters $\theta_2, \theta_3, \theta_4$, that the value 0 is inside the confidence intervals. Obviously, the hypothesis "$\theta_2 = \theta_3 = \theta_4 = 0$" is meaningless. Figure 2.4 illustrates the discrepancy between the distribution of $\widehat{\theta} - \theta$ and its approximation by the distribution $\mathcal{N}(0, V_\theta)$. In this figure, the sections of the confidence ellipsoids and likelihood contours in the plane (θ_1, θ_2) clearly differ significantly in their appearance. Thus, we cannot use the percentiles of a Gaussian distribution to calculate the confidence intervals. Instead, let us try the bootstrap method.

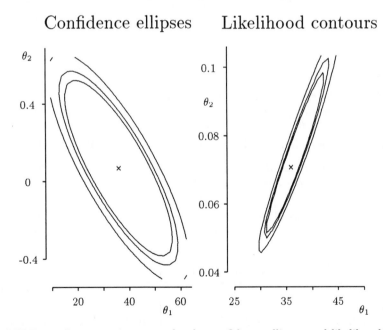

FIGURE 2.4. Isomerization example: the confidence ellipses and likelihood contours for the parameters (θ_1, θ_2) are drawn at levels 90%, 95% and 99%

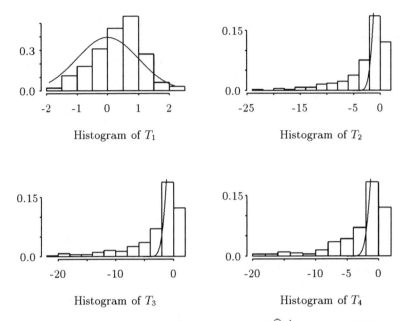

FIGURE 2.5. Isomerization example: histogram of $(\widehat{T}^{*,b}, b = 1, \ldots B)$ for each parameter. The line is the probability density of a $\mathcal{N}(0,1)$

Calculation of confidence intervals using bootstrap. $B = 199$ bootstrap simulations have been done to calculate another approximation of the distribution of $\widehat{\theta}$. Let

$$\widehat{T}_a = \frac{\widehat{\theta}_a - \theta_a}{\widehat{S}_a},$$

where θ_a is the component a of θ and $\widehat{S}_a$ is the estimation of the standard error of $\widehat{\theta}_a$. For each parameter, the bootstrap estimation of the distribution of $\widehat{T}_a$ is shown in figure 2.5. Note that the bootstrap distributions are very far from the Gaussian distribution, except for the first parameter.

Bootstrap estimations of standard error and bias for each parameter estimator. The results are shown table 2.4.

TABLE 2.4. Isomerization example. Bootstrap estimation of standard errors and bias for the parameters θ. 2.5% and 97.5% percentile of the bootstrap distribution of $\widehat{T}$. Bootstrap confidence intervals for the θ

	$\widehat{S}^*$	$\widehat{\mathrm{BIAS}}^*$ (percentage of bias)	$b_{0.025}$	$b_{0.975}$	$\widehat{I}_B$
θ_1	9.83	5.47 (15%)	-1.38	1.69	[23.2 46.2]
θ_2	0.133	0.002 (3%)	-15.2	0.137	[0.048 2.56]
θ_3	0.080	0.002 (6%)	-16.4	0.151	[0.024 1.54]
θ_4	0.322	0.008 (5%)	-15.4	0.144	[0.112 6.05]

TABLE 2.5. Isomerization example with the new parameterization. Estimated parameters and standard errors. Normal confidence intervals

	Estimates	Standard errors ($\widehat{S}$)	95% confidence interval $\widehat{I}_{\mathcal{N}}$
β_1	0.73738	1.66	[-2.51 3.98]
β_2	0.052274	0.00418	[0.0441 0.0605]
β_3	0.027841	0.00581	[0.0164 0.0392]
β_4	0.12331	0.0161	[0.0917 0.155]
σ^2	0.13477		

The bootstrap estimations of the standard errors, $\widehat{S}_a^*$ (see equation (2.8)), are of the same magnitude as $\widehat{S}_a$. The bootstrap yields a high value of the bias (see equation (2.9)) for θ_1, but the bias is small for the other parameters. The 0.025 and 0.975 percentiles of the $\widehat{T}_a^{*,b}$ ($\widehat{T}_a^*$ is the bootstrap version of $\widehat{T}_a$) are calculated as in section 2.4.1. They show the asymmetry of the estimators' distribution. By comparison, the 0.025 percentile of a Gaussian $\mathcal{N}(0,1)$ distribution is $\nu_{0.975} = 1.96$. The last column gives the bootstrap confidence intervals (see equation 2.7). For the three last parameters the lower bound of the intervals is positive; this condition is more realistic than the negative bounds obtained with $\widehat{I}_{\mathcal{N}}$. The confidence intervals are not symmetric around the estimated values $\widehat{\theta}_a$.

Calculation of confidence intervals using a new parameterization of the function f. An alternative to the bootstrap is to find another parameterization of the function $f(x,\theta)$ that reduces the discrepancy between the distribution of $\widehat{T}$ and the approximation of $\widehat{T}$ by a Gaussian distribution.

Model. A new parameterization, already suggested by several authors (see [BW88] for example) is defined by considering a new set of parameters, say $(\beta_1, \beta_2, \beta_3, \beta_4)$. These are obtained by eliminating the product $\theta_1\theta_3$ in equation (1.9). The model function is now defined by

$$f(x,\beta) = \frac{P - I/1.632}{\beta_1 + \beta_2 H + \beta_3 P + \beta_4 I}, \tag{2.10}$$

and $\mathrm{Var}(\varepsilon_i) = \sigma^2$.

Parameter and standard error estimations. The results are given in table 2.5. The estimated accuracy of the parameters is reasonable, and the discrepancy between confidence ellipses and likelihood contours for the pair (β_1, β_2) is not as big as it is for the pair (θ_1, θ_2); see figure 2.6.

Let us assume that we are interested in calculating confidence intervals for the parameters θ. We want to calculate confidence intervals for each θ_a by using the confidence regions calculated for β. In our example, the relations between the θ and β are easy to write:

$$\theta_1 = 1/\beta_3$$

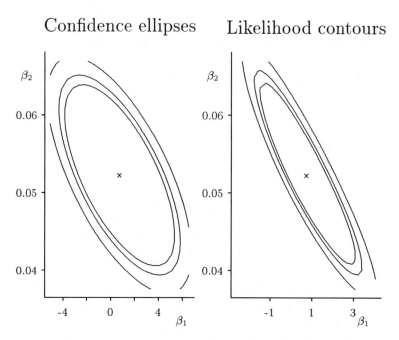

FIGURE 2.6. Isomerization example: the confidence ellipses and likelihood contours for the parameters (β_1, β_2) are drawn at levels 90%, 95% and 99%

$$\begin{aligned} \theta_2 &= \beta_2/\beta_1 \\ \theta_3 &= \beta_3/\beta_1 \\ \theta_4 &= \beta_4/\beta_1. \end{aligned}$$

If β_3 lies in the interval $[0.0164, 0.0392]$, then θ_1 lies in $[25.5, 60.8]$; however, we see that the same reasoning cannot be applied for θ_2, because 0 lies in the confidence interval for β_1. Thus, in this example, the calculation of a confidence interval for the parameters θ_a, $a = 1, \ldots p$ is only possible for θ_1. Only the bootstrap method allows us to calculate confidence intervals for each parameter.

2.4.6 Pasture regrowth: calculation of a confidence interval for $\lambda = \exp \theta_3$

Let us return to our pasture regrowth example to illustrate how to calculate a confidence interval that respects the constraint that parameter λ must be positive.

Model. The regression function is

$$f(x, \theta) = \theta_1 - \theta_2 \exp\left(-\exp(\theta_3 + \theta_4 \log x)\right),$$

and the variances are homogeneous: $\text{Var}(\varepsilon_i) = \sigma^2$.

Results.

Parameters	Estimated values	Asymptotic covariance matrix			
θ_1	69.95	3.09			
θ_2	61.68	3.87	6.66		
θ_3	-9.209	0.76	1.25	0.37	
θ_4	2.378	-0.22	-0.35	-0.09	0.027
σ^2	0.9306				

The parameter of interest is $\lambda(\theta) = \exp \theta_3$.

Calculation of confidence intervals with asymptotic level 95%, using results of section 2.3.2.

$\widehat{\lambda}$	$\widehat{S}$	$\nu_{0.975}$	$\widehat{I}_{\mathcal{N}}$
0.0001001	0.0000609	1.96	[-.0000194 , .000219]

It is immediately apparent that this confidence interval is unusable because λ cannot be negative. The result would have been the same if we had estimated λ by replacing the model function (1.1) with the following one: $f(x, \theta_1, \theta_2, \lambda, \theta_4) = \theta_1 - \theta_2 \exp(-\lambda x^{\theta_4})$.

Another way to calculate a confidence interval for λ is to transform the confidence interval calculated for θ_3. If θ_3 lies in $[-10.4 - 8.01]$, then $\lambda = \exp \theta_3$ lies in $[.0000303 \ .000331]$. By construction, this confidence interval is adapted to the set of variation of λ.

2.5   Conclusion

We have proposed several methods of estimating confidence intervals based on approximating the distribution of $\widehat{\lambda}$ either by the Gaussian distribution or by the bootstrap distribution. In some cases (see example 2.4.1), it is more convenient to consider a monotone transformation of the parameter of interest in place of the parameter itself, because the distribution of its estimator is better approximated by a Gaussian distribution. In other cases, the bootstrap method is more appropriate (see example 2.4.5). The examples treated above show that there is no rule for deciding in advance which is the "right" method. Nevertheless, in each case, the final choice was based on the adequacy of the result and the nature of the parameter of interest.

2.6   Using **nls2**

This section reproduces the commands and files used in this chapter to analyze the examples using **nls2**. It is assumed that the commands introduced in section 1.6 have already been executed.

Pasture regrowth example: confidence interval for the maximum yield

The results of estimation have been stored in the structure `pasture.nl1` (see section 1.6, page 20). Now we want to calculate a confidence interval for the maximum yield, the parameter θ_1.

Confidence interval for $\lambda = \theta_1$ with asymptotic level 95%

We use the function `confidence`. This function calculates the confidence interval $\widehat{I}_N$ defined by equation (2.2), page 32, and the confidence interval $\widehat{I}_T$ defined by equation (2.3), page 33. By default, the asymptotic level is 95%.

```
> pasture.conf.par <- confidence(pasture.nl1)
```

We display the values of $\widehat{\lambda}$, $\widehat{S}$, $\nu_{0.975}$, $\widehat{I}_N$, $t_{0.975}$ (5 degrees of freedom) and $\widehat{I}_T$:

```
> cat("Estimated value of lambda:", pasture.conf.par$psi[1],"\n" )
> cat("Estimated value of S:",pasture.conf.par$std.error[1],"\n" )
> cat("nu_(0.975):", qnorm(0.975),"\n" )
> cat("Estimated value of In:",
      pasture.conf.par$normal.conf.int[1,],"\n" )
> cat("t_(0.975, 5):", qt(0.975,5),"\n" )
> cat("Estimated value of It:",
      pasture.conf.par$student.conf.int[1,],"\n" )
```

Confidence interval for $\lambda = \theta_1$ using bootstrap with asymptotic level 95%

To calculate the confidence interval $\widehat{I}_B$ defined by equation (2.7), page 37, we use the function `bootstrap`. Several methods of bootstrap simulations are possible. Here, we choose `residuals`, which means that pseudoerrors are randomly generated among the centered residuals.

To initialize the iterative process of `bootstrap`, `nls2` must be called with the option `renls2`. We also set the option `control` so that intermediary results are not printed. Finally, we call the function `delnls2` to destroy any internal structures:

```
> pasture.nl1 <- nls2(pasture, "pasture.mod1",
          list(theta.start= c(70, 60, 0, 1), max.iters=100),
          control=list(freq=0),
          renls2=T)
> pasture.boot <- bootstrap(pasture.nl1,
            method="residuals",
            n.loops=199)
> delnls2()
```

We calculate the values of $\widehat{T}^{*,b}$, $b = 1, \ldots 199$ (see section 2.3.5) and illustrate their distribution function by plotting them in a histogram (see figure 2.1, page 39):

```
> P1.B <- pasture.boot$pStar[,1]
> SE.P1.B <- sqrt(pasture.boot$var.pStar[,1])
> T.B <- (P1.B-pasture.nl1$theta[1])/SE.P1.B
> hist(T.B,nclass=12,
        title="Pasture regrowth example",
        sub="Histogram of bootstrap estimations for T")
```

We sort the values of T.B to calculate the 0.0025 and 0.975 percentiles of the $\widehat{T}^{*,b}$, $b_{0.025}$ and $b_{0.975}$ (see section 2.4.1), and we display the values of $\widehat{\lambda}$, $\widehat{S}$, $b_{0.025}$, $b_{0.975}$ and $\widehat{I}_B$:

```
> b0 _ sort(T.B)
>    # Print the results:
> cat("Estimated value of lambda:", pasture.nl1$theta[1],"\n")
> cat("Estimated value of S:",coef(pasture.nl1)$std.error[1],"\n")
> cat("b_(0.025):", b0[5],"\n" )
> cat("b_(0.975):", b0[195],"\n" )
> cat("Estimated value of Ib:",
    pasture.nl1$theta[1]+b0[5]*coef(pasture.nl1)$std.error[1],
    pasture.nl1$theta[1]+b0[195]*coef(pasture.nl1)$std.error[1],"\n")
```

Finally, we calculate the accuracy characteristics of the bootstrap estimation: the bias $(\widehat{\mathrm{BIAS}}^*$, equation (2.9)), the variance $(\widehat{S}^*$, equation (2.8)), the mean square error $(\widehat{\mathrm{MSE}}^*)$, and the median $(\widehat{\mathrm{MED}}^*)$:

```
> cat("BIAS:" , (mean(P1.B)-pasture.nl1$theta[1]),"\n" )
> cat("S:" ,sqrt(var(P1.B)),"\n" )
> cat("MSE:" ,
      var(P1.B)+(mean(P1.B)-pasture.nl1$theta[1])^2 ,"\n" )
> cat("MED:" ,median(P1.B) ,"\n" )
```

Note: The bootstrap method generates different numbers at each execution. Thus, results of these commands may vary slightly from those displayed in section 2.4.1, page 38.

Cortisol assay example: confidence interval for D

We calculated an estimate of the calibration curve (see section 1.6, page 21) in structure corti.nl1, and now we want to calculate a confidence interval for the estimation of the dose of hormone D contained in a preparation that has the expected response $\mu = 2000$ c.p.m.

Confidence interval for D

We describe the function D in a file called corti.D. The expected response μ is introduced by the key word **pbispsi**:

```
psi D;
ppsi n,d,a,b,g;
pbispsi mu;
aux X1, X2, X;
```

```
subroutine;
begin
X1 = log((d-n)/(mu-n));
X2 = exp(X1/g);
X = (log(X2-1)-a)/b;
D = 10**X;
end
```

To calculate a confidence interval for D, we apply the confidence function. Then, we display the results of interest: $\widehat{D}$, $\widehat{S}$, $\nu_{0.975}$ and $\widehat{I}_{\mathcal{N}}$:

```
> corti.conf.D <- confidence(corti.nl1,file="corti.D",pbispsi=2000)
>    # Print the results:
> cat("Estimated value of D:", corti.conf.D$psi,"\n" )
> cat("Estimated value of S:",corti.conf.D$std.error,"\n" )
> cat("nu_(0.975):", qnorm(0.975),"\n" )
> cat("Estimated value of In:",corti.conf.D$normal.conf.int,"\n" )
```

(The results are given in section 2.4.2, page 40.)

ELISA test example: comparison of curves

The results of estimation have been stored in the structure elisa.nl1 (see section 1.6, page 23). Now we want to test the *parallelism* of the May and June curves.

Wald test with asymptotic level 95%

To test the parallelism of the response curves by a Wald test (see section 2.3.3), we use the function wald.

First we describe the functions to be tested in a file called elisa.wald:

```
psi d1,d2,d3;
ppsi p1_c1, p2_c1, p3_c1, p1_c2, p2_c2, p3_c2;
subroutine;
begin
d1=p1_c1-p1_c2;
d2=p2_c1-p2_c2;
d3=p3_c1-p3_c2;
end
```

We apply the function wald, and we display the value of the statistic S_W and the 0.95 quantile of a χ^2 with 3 degrees of freedom from which it should be compared:

```
> elisa.wald <- wald(elisa.nl1,file="elisa.wald")
>    # Print the results:
> cat("SW:",elisa.wald$statistic,"\n" )
> cat("X2(3):", qchisq(0.95, 3),"\n" )
```

Because the variances are homogeneous, we calculate the test statistic defined in equation (2.5), page 35:

```
> SF <- (elisa.wald$statistic*(32-8))/(32*3)
> cat("SF:", SF,"\n" )
> cat("F(3,24):", qf(0.95, 3,24), "\n" )
```

Likelihood ratio tests

To test the parallelism of the curves by likelihood ratio tests (see section 2.3.3), we have to estimate the parameters under hypothesis A (*the parallelism of the curves is not verified*), under hypothesis H (*the curves are parallel*) and under the last hypothesis (*the curves are identical*).

Estimation under hypothesis A has already been calculated: A is the hypothesis under which structure `elisa.nl1` has been built.

Estimation under hypothesis H is done by setting equality constraints on all of the parameters, except for the last one. Equality constraints are given with the option `eqp.theta`:

```
> elisa.nlH <- nls2(elisa,
    list(file="elisa.mod1", eqp.theta=c(1,2,3,4,1,2,3,5)),
    rep(c(2,0,1,0),2))
```

Estimation under the last hypothesis is done by setting equality constraints on all of the parameters:

```
> elisa.nlb <- nls2(elisa,
    list(file="elisa.mod1", eqp.theta=c(1,2,3,4,1,2,3,4)),
    rep(c(2,0,1,0),2))
```

We display the estimated values of the parameters and the sums of squares for the three hypothesis:

```
> cat("Estimated values of the parameters for the 3 hypothesis:\n")
> print(elisa.nl1$theta)
> print(elisa.nlH$theta)
> print(elisa.nlb$theta)
> cat("Estimated sums of squares for the 3 hypothesis:\n",
    elisa.nl1$rss, "\n", elisa.nlH$rss,"\n",
    elisa.nlb$rss,"\n")
```

Now, we calculate the test statistic S_L and display the 0.95 quantile of a χ^2 with 1 degree of freedom from which they should be compared. We also print the estimated value of $\beta = (\theta_4^{May} - \theta_4^{June})$:

```
> cat("Sl:",
    32*(log(elisa.nlH$rss) - log(elisa.nl1$rss)),
    32*(log(elisa.nlb$rss) - log(elisa.nlH$rss)),"\n" )
> cat("X2(0.95,1):", qchisq(0.95,1),"\n" )
> cat("Estimated value of beta:",
    elisa.nlH$theta["p4_c1"] -  elisa.nlH$theta["p4_c2"],"\n" )
```

Confidence interval for ρ with asymptotic level 95%

Now, we want to calculate a confidence interval for a function of the parameters: $\rho = \lambda(\theta) = 10^{(\theta_4^{June} - \theta_4^{May})}$.

We describe ρ in a file called `elisa.ro`:

```
psi ro;
ppsi p4_c1, p4_c2;
subroutine;
begin
ro=10**(p4_c2-p4_c1);
end
```

The function `confidence` is applied to the structure `elisa.nlH`, which contains the results of estimation under hypothesis H (*the curves are parallel*). We display the values of $\widehat{\rho}$, the standard error $(\widehat{S})$, $\nu_{0.975}$, $\widehat{I}_{\mathcal{N}}$, $t_{0.975}$ (27 degrees of freedom) and $\widehat{I}_T$:

```
>  elisa.ro <- confidence(elisa.nlH, file="elisa.ro")
>     # Print the results:
> cat("Estimated value of rho:", elisa.ro$psi,"\n" )
> cat("Estimated value of S:",elisa.ro$std.error,"\n" )
> cat("nu_(0.975):", qnorm(0.975),"\n" )
> cat("Estimated value of In:",elisa.ro$normal.conf.int,"\n" )
> cat("t_(0.975, 27):", qt(0.975,27),"\n" )
> cat("Estimated value of It:",elisa.ro$student.conf.int,"\n" )
```

Confidence interval for ρ using bootstrap with asymptotic level 95%

To calculate confidence intervals for ρ with bootstrap simulations (see section 2.3.5, page 35) we apply the function `bootstrap`.

To initialize the iterative bootstrap process, `nls2` is first called with the option `renls2`, and, finally, the function `delnls2` cleans the internal structures:

```
> elisa.nlH <- nls2(elisa,
    list(file="elisa.mod1", eqp.theta=c(1,2,3,4,1,2,3,5)),
    rep(c(2,0,1,0),2),
    control=list(freq=0),
    renls2=T)
> elisa.boot.ro <- bootstrap(elisa.nlH,method="residuals",
                    file="elisa.ro", n.loops=199)
> delnls2()
```

We display the values of $\widehat{\rho}$, $\widehat{S}$, $b_{0.025}$, $b_{0.975}$ and $\widehat{I}_B$:

```
> cat("Estimated value of rho:",elisa.ro$psi,"\n" )
> cat("Estimated value of S:", elisa.ro$std.error,"\n" )
> cat("b_(0.025):", sort(elisa.boot.ro$tStar)[5],"\n" )
> cat("b_(0.975):", sort(elisa.boot.ro$tStar)[195],"\n" )
> cat("Estimated value of Ib:", elisa.boot.ro$conf.int ,"\n" )
> cat("Bootstrap standard error:", sqrt(var(elisa.boot.ro$psiStar))
```

To illustrate the distribution function of $\widehat{T}^*$, we plot a histogram of their values (see figure 2.2, page 43).

```
> hist(elisa.boot.ro$tStar, nclass=9,
```

```
           title="ELISA example",
           sub="Histogram of bootstrap estimations for T")
```

Note: The bootstrap method generates different numbers at each execution. Thus, results of these commands may vary slightly from those displayed in section 2.4.3, page 40.

Ovocytes example

Confidence ellipsoids and likelihood contours for the parameters (P_w, P_s)

The results of estimation by nls2 have been stored in the structure ovo.nl1 (see section 1.6, page 25). We want to compare the parameters P_w and P_s by calculating confidence ellipsoids and likelihood contours in the space of these parameters.

The functions ellips and iso are used. ellips returns what is necessary to plot confidence ellipsoids, and iso returns what is necessary to define confidence regions in a two-dimensional space of parameters. The plots themselves are drawn by the graphical functions of **S-PLUS**:

```
> ovo.ell1 <- ellips(ovo.nl1, axis=c("Pw_c1","Ps_c1"))
> ovo.ell2 <- ellips(ovo.nl1, axis=c("Pw_c2","Ps_c2"))
> ovo.iso1 <- iso(ovo.nl1, axis=c("Pw_c1","Ps_c1"))
> ovo.iso2 <- iso(ovo.nl1, axis=c("Pw_c2","Ps_c2"))
>    # Graphical functions of Splus
> par(mfrow=c(1,2))
> plot(x=c(.06,.13),y=c(0.0008,.0017),type="n",xlab="Pw",ylab="Ps")
> contour(ovo.ell1,levels=qchisq(0.95,2),add=T,labex=0)
> contour(ovo.ell2,levels=qchisq(0.95,2),add=T,labex=0)
> text(0.1,0.0015,"mature ovocytes")
> text(0.08,0.001,"immature ovocytes")
> title("Confidence ellipses")
> plot(x=c(.06,.13),y=c(0.0008,.0017),type="n",xlab="Pw",ylab="Ps")
> contour(ovo.iso1,levels=qchisq(0.95,2),add=T,labex=0)
> contour(ovo.iso2,levels=qchisq(0.95,2),add=T,labex=0)
> text(0.095,0.0015,"mature ovocytes")
> text(0.08,0.001,"immature ovocytes")
> title("Likelihood contours")
```

(See figure 2.3, page 44).

Isomerization example

We have calculated one estimate of the parameters (see section 1.6, page 27) in the structure isomer.nl1, and now we want to calculate confidence intervals for each parameter.

Confidence intervals for each parameter with asymptotic level 95%

We use the function confidence to calculate the confidence interval $\widehat{I}_N$ defined by equation (2.2), page 32, for each parameter:

```
> isomer.conf.par <- confidence(isomer.nl1)
```

We display the estimated values of the parameters, their standard errors $(\widehat{S})$ and the confidence interval $\widehat{I}_N$:

```
> print(matrix(c(isomer.conf.par$psi,
                 isomer.conf.par$std.error,
                 isomer.conf.par$normal.conf.int[,"lower"],
                 isomer.conf.par$normal.conf.int[,"upper"]),
            ncol=4,
            dimnames=list(names(isomer.conf.par$psi),
              c("parameters","std","lower bound","upper bound" ))))
```

(Results are shown table 2.3, page 45.)

Confidence regions for parameters

We use the functions `ellips` and `iso` and the graphical functions of **S-PLUS** to plot confidence ellipsoids and likelihood contours in the space of the parameters (θ_1, θ_2):

```
> isomer.ell <- ellips(isomer.nl1, axis=c(1,2))
> isomer.cont <- iso(isomer.nl1, axis=c(1,2),
              bounds=matrix(c(25,50,0.03,0.11),nrow=2))
>    # Graphical functions of Splus
> par(mfrow=c(1,2))
> contour(isomer.ell,levels=qchisq(c(0.90,0.95,0.99),4),labex=0)
> points(x=isomer.nl1$theta[1], y=isomer.nl1$theta[2])
> title("Confidence ellipses")
> contour(isomer.cont,levels=qchisq(c(0.90,0.95,0.99),4),labex=0)
> title("Likelihood contours")
> points(x=isomer.nl1$theta[1], y=isomer.nl1$theta[2])
```

(See figure 2.4, page 45).

Calculation of confidence intervals using bootstrap

Confidence intervals using the bootstrap method (see section 2.3.5, page 35) are calculated by using the function `bootstrap`.

Here, to reduce the execution time, which may be long because the model must be calculated several times at each loop, we choose to evaluate by the C-program rather than by syntaxical trees (see Ovocytes example, section 1.6, page 25).

To generate and compile the program that calculates the model, we type the operating system commands:

```
$  analDer isomer.mod1
$  Splus COMPILE isomer.mod1.c INC=$SHOME/library/nls2/include
```

We then load the program into our **S-PLUS** session:

```
> loadnls2("isomer.mod1.o")
```

To initialize the iterative bootstrap process, nls2 first is called with the option renls2, and, finally, the function delnls2 cleans the internal structures:

```
> isomer.nl1 <- nls2(isomer,"isomer.mod1", c(36,.07,.04,.2),
            control=list(freq=0), renls2=T)
> isomer.boot <- bootstrap(isomer.nl1,
            method="residuals", n.loops=199)
> delnls2()
```

Histograms of $(\widehat{T}^{\star,b}, b = 1,\ldots 199)$ for each parameter

Histograms of $\widehat{T}^\star$ for each parameter illustrate the boostrap estimation of their distribution. Only the results corresponding to correct estimations, i.e. when isomer.boot$code=0, are taken into account:

```
> pStar_ isomer.boot$pStar[isomer.boot$code==0,]
> var.pStar_ isomer.boot$var.pStar[isomer.boot$code==0,]
> theta<-matrix(rep(isomer.nl1$theta,isomer.boot$n.loops),
            ncol=4, byrow=T)
> TT <-(pStar - theta)/  sqrt(var.pStar)
> par(mfrow=c(2,2))
> for (a in 1:4)
> {
> hist(TT[,a],probability=T,main="Isomerization example",
      sub=paste("Histogram of bootstrap estimations for T",a),xlab="")
> qx<-seq(from=min(TT[,a]),to=max(TT[,a]),length=75)
> lines(qx,dnorm(qx))
> }
```

(See figure 2.5, page 46).

Bootstrap estimations of standard error and bias for each parameter estimator

We calculate the accuracy characteristics of the bootstrap estimation and display the values of the standard error $(\widehat{S}^\star)$, the bias $(\widehat{BIAS}^\star)$, the percentage of bias, the 0.0025 and 0.975 percentiles ($b_{0.025}$ and $b_{0.975}$) and the confidence interval $\widehat{I}_B$:

```
> SE.boot <- sqrt(diag(var(pStar)))
> bias.boot <-  apply(pStar,2,mean)-isomer.nl1$theta
> Pbias.boot <- 100*bias.boot/isomer.nl1$theta
> sortedTT_apply(TT,2,sort)
> B_seq(1:isomer.boot$n.loops)/isomer.boot$n.loops
> i025_1
> while (B[i025] < 0.25) { i025_i025+1 }
> i975_isomer.boot$n.loops
> while (B[i975] >= 0.975) { i975_i975-1 }
> b0.025.boot <- sortedTT[i025,]
> b0.975.boot <- sortedTT[i975+1,]
> binf.boot <- isomer.nl1$theta -
            b0.975.boot*coef(isomer.nl1)$std.error
```

```
> bsup.boot <- isomer.nl1$theta -
        b0.025.boot*coef(isomer.nl1)$std.error
>    # Print the results:
> print(matrix(c(SE.boot, bias.boot,Pbias.boot,
                 b0.025.boot,b0.975.boot,
                 binf.boot, bsup.boot), ncol=7,
          dimnames=list(names(isomer.nl1$theta),
                c("S","BIAS","% of BIAS","b0.025","b0.975",
                  "lower bound","upper bound" ))))
```

Note: The bootstrap method generates different numbers at each execution. Thus, results of these commands may vary slightly from those displayed in section 2.4, page 46.

Confidence intervals using a new parameterization of the function f

A new parameterization of the regression function f now is considered. The model of equation (2.10), page 47, is defined in a file called isomer.mod2:

```
resp r;
varind H,P,I;
aux a1, a2;
parresp b1,b2,b3,b4;
subroutine;
begin
a1= P - I/1.632;
a2= b1 + b2*H + b3*P + b4*I;
r=a1/a2;
end
```

Before calling nls2 to estimate the parameters, we have to call the function loadnls2. If we do not do this, the program isomer.mod1.o, just previously loaded into the **S-PLUS** session, will still be current. loadnls2 is called without any argument to reset the default action; the default is to calculate the model by syntaxical trees:

```
> loadnls2()
>   isomer.nl2<-nls2(isomer,"isomer.mod2",rep(1,4))
```

Confidence intervals are calculated by using the function confidence.
   We display the estimated values of the parameters, their standard errors $(\widehat{S})$, and the 95% confidence interval $\widehat{I}_{\mathcal{N}}$:

```
> isomer.conf.par2 <- confidence(isomer.nl2)
> print(matrix(c(isomer.conf.par2$psi,
                 isomer.conf.par2$std.error,
                 isomer.conf.par2$normal.conf.int[,"lower"],
                 isomer.conf.par2$normal.conf.int[,"upper"]),
          ncol=4,
          dimnames=list(names(isomer.conf.par2$psi),
                c("parameters","S",
                  "lower bound","upper bound" ))))
```

(Results are given in table 2.5, page 47.)

Confidence regions with the new parametrization

We plot confidence ellipses and likelihood contours in the space of the
parameters (β_1, β_2) by using the functions `ellips` and `iso` and graphical
functions of **S-PLUS**:

```
> isomer.ell2 <- ellips(isomer.nl2, axis=c(1,2))
> isomer.iso2 <- iso(isomer.nl2, axis=c(1,2))
>    # Graphical functions of Splus
> par(mfrow=c(1,2))
> plot(x=c(-5,7),y=c(0.03,.07),type="n",xlab="b1",ylab="b2")
> contour(isomer.ell2,levels=qchisq(c(0.90,0.95,0.99),4),
          add=T,labex=0)
> points(x=isomer.nl2$theta[1], y=isomer.nl2$theta[2])
> title("Confidence ellipses")
> plot(x=c(-5,7),y=c(0.03,.07),type="n",xlab="b1",ylab="b2")
> contour(isomer.iso2,levels=qchisq(c(0.90,0.95,0.99),4),
          add=T,labex=0)
> title("Likelihood contours")
```

(See figure 2.6, page 48).

Pasture regrowth example: confidence interval for $\lambda = \exp\theta_3$

Let us return to the pasture regrowth example to calculate a confidence
interval for $\lambda = \exp\theta_3$.

We define the function λ in a file called `pasture.lambda`:

```
psi lambda;
ppsi p3;
subroutine;
begin
lambda = exp(p3);
end
```

A confidence interval for λ is calculated by using the `confidence` func-
tion. We display the values of $\widehat{\lambda}$, $\widehat{S}$, $\nu_{0.975}$ and $\widehat{I}_\mathcal{N}$:

```
> pasture.conf.expP3 <- confidence(pasture.nl1,
              file="pasture.lambda")
>    # Print the results:
> cat("Estimated value of lambda:", pasture.conf.expP3$psi,"\n" )
> cat("Estimated value of S:",pasture.conf.expP3$std.error,"\n" )
> cat("nu_(0.975):", qnorm(0.975),"\n" )
> cat("Estimated value of In for exp(p3):",
    pasture.conf.expP3$normal.conf.int,"\n" )
> cat("Estimated value of In for p3:",
    pasture.conf.par$normal.conf.int[3,],"\n" )
> cat("Exponential transformation of the preceding interval:",
    exp(pasture.conf.par$normal.conf.int[3,]),"\n" )
```

(Results for this example are given in section 2.4.6, page 48.)

3

Variance estimation

In the radioimmunological assay of cortisol example, we introduced the necessity of using nonlinear regression models with heterogeneous variances, and we suggested the weighted least squares method for analyzing this particular data set (see sections 1.1.2, 1.3 and 2.4.2). This method, however, is not adequate for every situation. Whereas the radioimmunological assay of cortisol data provided many replications for each variance, some data sets only provide a few replications, and some provide none at all, as we illustrate in the examples below.

Although one can feasibly use the weighted least squares method for few replications, it is not the best option because the accuracy of the empirical variance as an estimator of the true variance is rather bad: with 4 replications, the relative error is roughly 80%. The weighting by such inaccurate estimators can be very misleading.

To handle these situations with few or no replications and to still account for the heterogeneity of the variances, we present two alternative methods in this chapter: the maximum likelihood method and the three-step alternate mean squares method. Both of these are based on a parametric modeling of the variance.

Other methods are also available to solve the estimation problem for nonlinear heteroscedastic regressions. For example, R. Carroll and D. Ruppert in [CR88] propose to take into account heteroscedasticity and skewness by transforming both the data and the regression function.

TABLE 3.1. Data for the growth of winter wheat tillers

Sum of degree-days (base 0°C)	Dry matter weight (mg)			
405.65	113.386	90.500		
498.75	161.600	207.650		
567.25	309.514	246.743		
618.30	460.686	422.936		
681.45	1047.000	972.383	1072.022	1034.000
681.45	1169.767	1141.883	999.633	1266.290
681.45	868.662	1133.287		

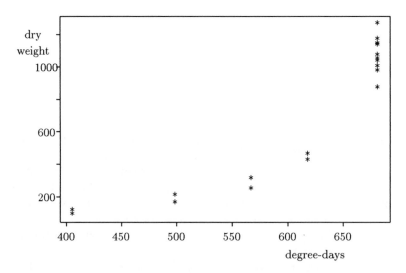

FIGURE 3.1. Observed dry weight of tillers on a degree days time scale

3.1 Examples

3.1.1 Growth of winter wheat tillers: few replications

In [FM88], the authors consider the growth of winter wheat, focusing on the differences in the dry weights of the wheat tillers, or stems. Time is measured on a cumulative degree-days scale with a 0°C base temperature and with a point of origin determined by the physiological state of the plants. Plants growing on randomly chosen small areas of about 0.15 m² are harvested each week. Table 3.1 and figure 3.1 document the tiller data, listing the means of the dry weights of the tillers for plants harvested from the same area.

The regression equation chosen to describe the increase in dry matter with respect to the cumulative sum of temperatures is the simple exponen-

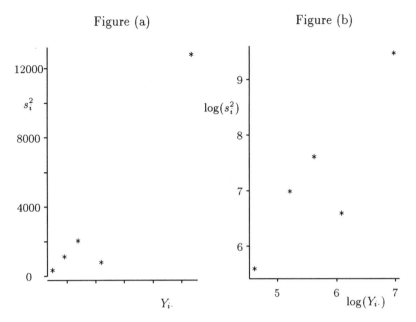

FIGURE 3.2. Winter wheat tillers example: figure (a) represents the empirical variances, s_i^2, versus mean for the tillers data $Y_i.$; figure (b) represents the same data after the logarithm transformation

tial function

$$f(x, \theta) = \theta_1 \exp(\theta_2 x) . \qquad (3.1)$$

Here, θ_2 is simply the relative growth rate. As $\exp(0) = 1$, θ_1 could be interpreted as the dry weight of the tillers at the origin; however, a precise physiological interpretation of θ_1 on this basis is risky.

Even though the design of the experiment is highly unbalanced, the variation of the variance of the dry weight with respect to the cumulative sum of temperatures is obvious. Figure 3.2 shows that the intrareplications variances vary with respect to the mean of the observations made at the same time.

Although the relative position of the data point corresponding to the fourth harvesting date is troublesome, we can represent the variance of the observations by an increasing function of the response function, $\sigma_i^2 = \sigma^2 f(x_i, \theta)$, for instance.

3.1.2   Solubility of peptides in trichloacetic acid solutions: no replications

In [CY92], the authors describe an experiment designed to investigate the precipitation mechanism of peptides. The percentage of solubility of 75 peptides, issued from the digestion of caseins, is related to their retention

TABLE 3.2. RP-HPLC retention time and solubility of 75 peptides

ret. time	% sol.	ret. time	% sol.	ret. time	% sol.
3.6	100.0	3.6	100.0	6.7	99.0
11.2	105.0	20.0	100.0	21.5	84.0
24.5	100.0	29.2	99.0	38.4	88.0
40.4	48.0	49.8	35.0	59.6	0.0
64.2	0.0	68.1	0.0	72.7	0.0
40.8	50.0	52.7	2.0	57.7	0.0
59.1	0.0	61.7	0.0	65.7	0.0
28.3	104.0	33.0	89.0	40.0	56.0
44.5	10.0	47.0	0.0	63.1	0.0
31.6	91.0	40.1	32.0	52.4	2.0
57.5	4.0	23.8	100.0	29.1	85.0
30.0	95.0	31.7	100.0	23.1	95.0
42.7	85.0	46.8	95.0	16.8	101.0
24.0	100.0	31.1	107.0	32.3	84.0
34.0	92.0	34.7	94.0	36.3	97.0
39.3	95.0	42.0	0.0	44.4	51.0
45.0	0.0	46.2	70.0	49.0	63.0
52.1	1.0	55.4	0.0	56.7	0.0
53.0	0.0	0.0	82.0	3.1	102.0
4.2	100.0	7.1	100.0	9.0	95.0
10.6	106.0	13.0	93.0	14.2	107.0
15.5	98.0	38.3	92.0	43.5	48.0
45.9	90.0	47.5	65.0	48.9	34.0
55.6	8.0	57.0	0.0	59.2	1.0
60.9	0.0	34.3	100.0	51.5	0.0

time on a column of a reverse-phase high-performance liquid chromatography device. The experiment is performed for various trichloacetic acid
concentrations; in table 3.2 and figure 3.3 we present the data, henceforth
called the peptide data, for only one of these concentrations.

The graph of the data presents the classical S shape, making the logistic function the natural candidate for the regression model. Moreover, as
percentages are considered, the asymptotes are, respectively, 0 and 100.
Further, it seems likely that these data have heterogeneous variances. No
replications are available to obtain estimates of the variance function for
each value of the independent variable; however, that the data are percentages suggests some binomial sampling scheme behind the observed phenomena. Hence, we can use a parabola to model the variance: when the
peptide is fully nonsoluble, its solubility can be assessed with a great pre-
cision, while the variability of the measure is maximum for intermediate
solubilities. The proposed model is as follows:

$$f(x, \theta) = \frac{100}{1 + \exp\left(\theta_2 \left(x - \theta_1\right)\right)} \tag{3.2}$$

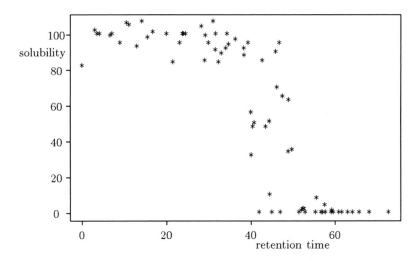

FIGURE 3.3. Peptides example: observed solubility of 75 peptides versus their RP-HPLC retention time

and

$$\sigma_i^2 = \sigma^2 + \sigma^2 \tau f(x_i, \theta)\left[100 - f(x_i, \theta)\right] = g\left(x_i, \theta, \sigma^2, \tau\right). \qquad (3.3)$$

When x varies from 0 to $+\infty$, $f(x, \theta)$ decreases from 100 to 0, whereas $g\left(x_i, \theta, \sigma^2, \tau\right)$ equals σ^2 when $f(x_i, \theta)$ equals 0, then increases and decreases and has a finite maximal value.

Actually, looking carefully at the data, it appears that the variability is greater when the peptide is fully soluble than when it is nonsoluble. This behavior is not taken into account by the preceding variance function, but this example will be discussed again in chapter 4.

3.2   Parametric modeling of the variance

Before we can estimate the variances, we must first choose a variance function. In some cases, this choice can be made on a more or less theoretical basis. But more often, we have only qualitative indications: either the variance of the observations grows with their expectation, or the relation between the variance of the observations and their expectation can be depicted by a parabola, for instance. In both cases, the choice of the model is based on the residuals inside a restricted collection of possible models.

For an increasing variance, there are essentially two models. Either the variance varies as a power of the response

$$\sigma_i^2 = g\left(x_i, \sigma^2, \theta, \tau\right) = \sigma^2 f(x_i, \theta)^\tau, \qquad (3.4)$$

or the variance varies as a linear function of the response

$$\sigma_i^2 = g\left(x_i, \sigma^2, \theta, \tau\right) = \sigma^2\left(1 + \tau f(x_i, \theta)\right). \qquad (3.5)$$

For a variance function varying like a parabola, we generalize the model given by equation (3.3). Let y_{min} be the smallest value of the regression function and y_{max} the biggest one. In general, y_{min} and y_{max} depend on θ. The model is then

$$\sigma_i^2 = g\left(x_i, \sigma^2, \theta, \tau\right) = \sigma^2 + \sigma^2 \tau_1 \left(y_{max} + \tau_2 - f(x_i, \theta)\right)\left(f(x_i, \theta) - y_{min}\right) .$$
(3.6)

Note that we have to be careful in using some of these models. A variance is positive by definition, but the above-mentioned functions can actually be negative for some values of the independent variable x and of the parameters. Let us consider the variance model given by equation (3.5) for instance. Even if the function $f(x, \theta)$ is nonnegative for any value of x and θ, the function $\sigma^2\left(1 + \tau f(x, \theta)\right)$ is negative when $\tau < -1/f(x, \theta)$.

3.3   estima-1

3.3.1   Maximum likelihood

The maximum likelihood method is widely known and used in the field of parametric inference. Statisticians like it because it behaves well in both theory and practice.

Let us recall the main features of the maximum likelihood method in the case of Gaussian observations. Assume that we observe a sample of independent outcomes from a Gaussian sampling scheme with unknown mean μ and variance σ^2. In this model, the probability density at point u is

$$\frac{1}{\sqrt{2\pi\sigma^2}} \exp\left(-\frac{(u - \mu)^2}{2\sigma^2}\right).$$

If X is an outcome from this sampling scheme, the likelihood is the random quantity

$$\frac{1}{\sqrt{2\pi\sigma^2}} \exp\left(-\frac{(X - \mu)^2}{2\sigma^2}\right).$$

For a sample of independent outcomes $X_1, \ldots, X_n$, the likelihood is the product of the likelihoods of each observation:

$$L_n(X_1, \ldots, X_n; \mu, \sigma^2) = \frac{1}{\left(\sqrt{2\pi\sigma^2}\right)^n} \prod_{i=1}^{n} \exp\left(-\frac{(X_i - \mu)^2}{2\sigma^2}\right) .$$

The maximum likelihood estimators for μ and σ^2 are those values that maximize the likelihood or, equivalently, its logarithm. In this case, they are simply the mean $\widehat{\mu} = n^{-1}\sum_{i=1}^{n} X_i$ and the empirical variance $\widehat{\sigma}^2 = n^{-1}\sum_{i=1}^{n}(X_i - \widehat{\mu})^2$.

Consider the nonlinear regression model

$$\left. \begin{array}{rcl} Y_i & = & f(x_i, \theta) + \varepsilon_i, \\ \text{Var}(\varepsilon_i) & = & g\left(x_i, \sigma^2, \theta, \tau\right), \ \text{E}(\varepsilon_i) = 0 \end{array} \right\}, \tag{3.7}$$

where the ε_i are assumed to be independent Gaussian random variables for i varying from 1 to n. For ease of handling, we generally use the logarithm of the likelihood (log-likelihood for short) instead of the likelihood itself. For the above heteroscedastic regression model, the log-likelihood

$$V(\theta, \sigma^2, \tau) = \log L_n(Y_1, \ldots Y_n, \theta, \sigma^2, \tau) \tag{3.8}$$

is given by the formula

$$V(\theta, \sigma^2, \tau) = -\frac{n}{2} \log 2\pi - \frac{1}{2} \sum_{i=1}^{n} \left[\log g\left(x_i, \sigma^2, \theta, \tau\right) + \frac{(Y_i - f(x_i, \theta))^2}{g\left(x_i, \sigma^2, \theta, \tau\right)} \right].$$

For a given set of observations $Y_1, \ldots, Y_n$, the log-likelihood is a function of the parameters. If p is the dimension of θ and q is the dimension of τ, the model depends on $p + q + 1$ parameters. The maximum likelihood estimator $\widehat{\theta}, \widehat{\sigma}^2, \widehat{\tau}$ maximizes the log-likelihood. The values of $\widehat{\theta}, \widehat{\sigma}^2, \widehat{\tau}$ cannot be given explicitly; they are obtained by numerical computation. Note also that if the variance of the observations is constant ($g\left(x, \sigma^2, \theta, \tau\right) = \sigma^2$ for each value of x), the maximum likelihood estimator and the ordinary least squares estimator are the same.

3.3.2 Three-step alternate mean squares

For some cases, such as model (3.7), the number of parameters is large, making the direct numerical maximization of the log-likelihood with respect to the complete set of parameters difficult. To overcome this difficulty, we can use the three-step alternate mean squares method. This method breaks down the estimation process into successive easier-to-handle steps, as we describe below.

1st step: ordinary least squares estimation of θ. A first estimator of θ is $\widehat{\theta}_{OLS}$, which minimizes $C(\theta) = \sum_{i=1}^{n} (Y_i - f(x_i, \theta))^2$ (see section 1.3).

2nd step: estimation of σ^2 and τ from the residuals by least squares. Because the ε_i are centered, their variance is also the expectation of their square. We compute the residuals $\widehat{\varepsilon}_{iOLS} = Y_i - f(x_i, \widehat{\theta}_{OLS})$. Then we estimate σ^2 and τ by ordinary least squares to solve the regression problem in which the observations are the $\widehat{\varepsilon}_{iOLS}^2$ and the regression function is $g(x, \sigma^2, \widehat{\theta}_{OLS}, \tau)$. The value of the parameter θ is assumed to be known and is given by its estimation from the preceding step. We denote by $\widetilde{\sigma}^2$ and $\widetilde{\tau}$ the values of the estimates.

3rd step: maximum likelihood estimation of θ. Coming back to the original data Y_i, we estimate θ using the maximum likelihood method and assume that the values of σ^2 and τ are known and are precisely $\widetilde{\sigma}^2$ and $\widetilde{\tau}$. More exactly, the estimator $\widehat{\theta}_{ALT}$ maximizes the function

$$-\frac{n}{2}\log 2\,\pi - \frac{1}{2}\sum_{i=1}^{n}\left[\log\left(g\left(x_i,\widetilde{\sigma}^2,\theta,\widetilde{\tau}\right)\right) + \frac{(Y_i - f(x_i,\theta))^2}{2\,g\left(x_i,\widetilde{\sigma}^2,\theta,\widetilde{\tau}\right)}\right].$$

This three-step alternate method has good properties: we can obtain classical asymptotic results analogous to the ones described in section 2.3.1. Moreover, it is of particular interest in case the user is faced with numerical difficulties, as can happen if reasonable initial values of the parameters are not easy to find out. However, based on our own experience and on the research, we strongly encourage the user to adopt the maximum likelihood approach as often as possible.

At the second and third steps of the three-step alternative method, other methods, called *modified least squares estimation* methods can also be used. These methods are derived from the weighted least squares method and will not be detailed in this book; see [HJM91] for more details. We will compare the maximum likelihood method with this modified least squares method in section 3.5.2.

3.4   Tests and confidence regions

3.4.1   The Wald test

As was just mentioned, the classical asymptotic results are available for the three-step alternate mean squares method. This is also true for the maximum likelihood method. As a consequence, for a continuous real function of the parameters $\lambda(\theta,\sigma^2,\tau)$, we have the following results. Let $\widehat{\theta},\widehat{\sigma}^2,\widehat{\tau}$ be the estimates of θ,σ^2,τ obtained by any one of both of the methods that we considered in the preceding section. Denote $\widehat{\lambda} = \lambda(\widehat{\theta},\widehat{\sigma}^2,\widehat{\tau})$ and $\lambda = \lambda(\theta,\sigma^2,\tau)$. Then $\widehat{\lambda} - \lambda$ tends to 0 as n tends to infinity. Moreover, there exists an asymptotic estimate $\widehat{S}$ of the standard error of $\widehat{\lambda}$ such that the distribution of $\widehat{S}^{-1}(\widehat{\lambda} - \lambda)$ can be approximated by a standard Gaussian distribution $\mathcal{N}(0,1)$. As in section 2.3.3, we can test the hypothesis H: $\{\lambda = \lambda_0\}$ against the alternative A: $\{\lambda \neq \lambda_0\}$ using the Wald statistic

$$\mathcal{S}_{\mathrm{W}} = \frac{(\widehat{\lambda} - \lambda_0)^2}{\widehat{S}^2}.$$

To this end, we use the following decision rule. We choose a risk level α, $\alpha < 1$; then we compute the critical value C such that $\Pr\{Z_1 > C\} = \alpha$, where Z_1 is a random variable distributed as a χ^2 with 1 degree of freedom; then we compare $\mathcal{S}_{\mathrm{W}}$ and C: if $\mathcal{S}_{\mathrm{W}} > C$, we reject the hypothesis H; if $\mathcal{S}_{\mathrm{W}} \leq C$, we do not reject it. This test has an asymptotic level α.

Remarks.

1. We have already seen various examples for λ in a homoscedastic regression model in section 2.4. So we can easily imagine what type of function of the parameters we can consider for inference in a heteroscedastic model, taking into account that the variance parameters (σ^2 and τ) also may appear in such a function. A very simple example is provided by $\lambda = \tau$ in the model given by equation (3.3) when analyzing the peptide data.

2. We have presented the Wald test without distinguishing between the maximum likelihood method and the three-step alternate mean squares method because the computations are formally the same. Obviously, because both methods are based on the fulfillment of distinct criteria (optimization of different functions), they do not give the same numerical results for a given set of observations. See section 3.5.2 for an illustration.

3. We have developed the Wald test only for a real (scalar) function of the parameters. This can be extended to r-dimensional functions of the parameters as in section 2.3.5, r being not greater than the total number of parameters.

3.4.2 The likelihood ratio test

In section 2.3.3, the likelihood ratio test statistic $\mathcal{S}_L$ was introduced for a regression model with homogeneous variances, the estimation method being the ordinary least squares. In that case, the computation of the statistic $\mathcal{S}_L$ was a particular case application of the general inference methodology based on the maximum likelihood principle. In this section, we describe how this principle applies when we are faced with a heteroscedastic regression model.

In order to treat this point in some generality, let us introduce some notations. We call ϕ the whole set of parameters $\phi^T = (\theta^T \sigma^2 \tau^T)$. We call q the dimension of τ, so ϕ is of dimension $p+q+1$. For instance, in the model described by equations (3.2) and (3.3) for the peptide data, $p = 2$, $q = 1$ and $\phi = (\theta_1, \theta_2, \sigma^2, \tau)^T$ is a 4-dimensional vector. In a lot of situations, $q = 0$, for instance if we describe the variance by $\sigma_i^2 = \sigma^2 f(x_i, \theta)$ as for the tiller data, or by $\sigma_i^2 = \sigma^2 f(x_i, \theta)^2$ as for the cortisol data.

We now consider a rather general situation where a hypothesis H is described by a linear constraint on a subset of the whole set of parameters ϕ. The constraint is expressed as $\Lambda\phi = L_0$, where Λ is an $r \times (p+q+1)$ matrix ($r < p+q+1$) and L_0 is a constant vector with dimension r. r is also assumed to be the rank of the matrix Λ. Let $\widehat{\phi}_H^T = (\widehat{\theta}_H^T, \widehat{\sigma}_H^2, \widehat{\tau}_H^T)$ be the parameter estimators for the hypothesis H, which means that the constraint $\Lambda\widehat{\phi}_H = L_0$ is fulfilled, whereas $\widehat{\theta}, \widehat{\sigma}^2, \widehat{\tau}$, the classical maximum likelihood estimators of the parameters of model (3.7), are the parameter estimators for the unconstrained alternative A. The likelihood ratio statistic is

$$\mathcal{S}_L = -2 \log \frac{L_n(Y_1, \ldots, Y_n; \widehat{\theta}, \widehat{\sigma}^2, \widehat{\tau})}{L_n(Y_1, \ldots, Y_n; \widehat{\theta}_H, \widehat{\sigma}_H^2, \widehat{\tau}_H)} . \qquad (3.9)$$

With our notations, this turns out to be

$$\mathcal{S}_L = \sum_{i=1}^{n} \log \frac{g(x_i, \widehat{\sigma}_H^2, \widehat{\theta}_H, \widehat{\tau}_H)}{g(x_i, \widehat{\sigma}^2, \widehat{\theta}, \widehat{\tau})} .$$

The hypothesis H is rejected if $\mathcal{S}_L > C$, where C is defined by

$$\Pr\{Z_r \le C\} = 1 - \alpha.$$

Z_r is distributed as a χ^2 with r degrees of freedom, and α is the (asymptotic) level of the test.

Remarks.

1. The intuitive justification of this decision rule is clear. On the one hand, if H is true, then $\widehat{\theta}_H, \widehat{\sigma}_H^2, \widehat{\tau}_H$ and $\widehat{\theta}, \widehat{\sigma}^2, \widehat{\tau}$ are convenient estimators of the model parameters and, therefore, take similar values; thus, the likelihood ratio is close to 1 and its logarithm is close to 0. On the other hand, if H is false, then $\widehat{\theta}_H, \widehat{\sigma}_H^2, \widehat{\tau}_H$ do not estimate the true value of the parameters and are expected to have values very different from $\widehat{\theta}, \widehat{\sigma}^2, \widehat{\tau}$; thus $\mathcal{S}_L$ has a value other than 0.

2. In section 2.3.3, the test statistic $\mathcal{S}_L$ is given by

$$\mathcal{S}_L = n \log C(\widehat{\theta}_H) - n \log C(\widehat{\theta}_A).$$

   This is the particular form taken by equation (3.9) in the homogeneous variance situation. Note that for an unconstrained alternative, $\widehat{\theta}_A = \widehat{\theta}$.

3.4.3   Links between testing procedures and confidence region computations

It is generally possible to compute a confidence region from a test procedure and *vice versa* to perform a test using a confidence region. For instance, going back to section 2.3, we show how the confidence interval $\widehat{I}_N$ given by equation (2.2) and the Wald test relate to each other. Actually, both are applications of the same result: the limiting distribution of $\widehat{T} = (\widehat{\lambda} - \lambda)/\widehat{S}$ is a standard normal distribution $\mathcal{N}(0, 1)$.

We can easily build a test procedure for the hypothesis H: $\{\lambda = \lambda_0\}$ against the alternative A: $\{\lambda \ne \lambda_0\}$ from the confidence interval $\widehat{I}_N$. The decision rule is simply reject H if the $\widehat{I}_N$ do not cover λ_0 and do not reject it if λ_0 is covered by $\widehat{I}_N$. It is not difficult to see that this test procedure exactly corresponds to the Wald test if we take the same asymptotic level α.

On the other hand, we can construct a confidence region for λ by considering the set of all of the values, say ℓ, of λ such that the Wald test does not reject the hypothesis $\{\lambda = \ell\}$. We define this confidence region $\widehat{I}_W$ as follows:

$$\widehat{I}_W = \left\{ \ell : \left(\widehat{\lambda} - \ell \right)^2 / \widehat{S}^2 \leq C \right\}.$$

Again, if C is such that $\Pr\{Z_1 \leq C\} = 1 - \alpha$, for Z_1 a χ^2 random variable with 1 degree of freedom, $\widehat{I}_W$ and $\widehat{I}_N$ do coincide exactly[1].

This duality between tests and confidence regions is quite general.

3.4.4 Confidence regions

Let us consider the peptide data and assume that the heteroscedastic regression model given by equations (3.2) and (3.3) is convenient. Suppose that we want to compute a confidence interval for the parameter θ_2. This parameter is linked to the maximal (in absolute value) slope of the logistic curve, which is $-25\,\theta_2$. Thus it is representative of the variability of the peptide solubility with respect to the retention time.

Let $\widehat{\theta}_1, \widehat{\theta}_2, \widehat{\sigma}^2, \widehat{\tau}$ be the maximum likelihood estimators of the parameters. Here, $\lambda(\theta, \sigma^2, \tau) = \theta_2$ and $\widehat{\lambda} = \widehat{\theta}_2$. $\widehat{S}$ is given simply by the square root of the second term of the diagonal of the matrix estimating the parameter estimates covariance. A confidence interval for θ_2 based on the asymptotic distribution of $\widehat{\theta}_2$ is then just $\widehat{I}_N$.

We also can, and this is what we recommend, compute a confidence interval for θ_2 using the maximum likelihood approach. This method is analogous to the profiling procedure described in [BW88]. To this end, we make use of the above-mentioned duality between tests and confidence regions. Consider first the test of the hypothesis "H: $\{\theta_2 = \ell\}$" against "A: $\{\theta_2 \neq \ell\}$" for a given real positive value ℓ. According to the notations introduced in section 3.4.2, the hypothesis H is described by the linear constraint

$$\begin{pmatrix} 0 & 1 & 0 & 0 \end{pmatrix} \begin{pmatrix} \theta_1 \\ \theta_2 \\ \sigma^2 \\ \tau \end{pmatrix} = \ell.$$

Here, $p = 2$, $q = 1$ and $r = 1$. The computation of the constrained estimators $\widehat{\theta}_{1\,H}, \widehat{\theta}_{2\,H}, \widehat{\sigma}^2_H, \widehat{\tau}_H$ is particularly simple in this case. We consider that

---

[1]Recall that C is such that, for $\alpha < 1/2$, $\sqrt{C}$ is the $1 - \alpha/2$ percentile of a standard Gaussian $\mathcal{N}(0,1)$ random variable.

θ_2 is no longer an unknown parameter, but a known constant equal to ℓ, and we maximize the likelihood with respect to θ_1, σ^2, τ. To emphasize the dependency on the value ℓ, we adopt another notation, writing $\theta_{1\,\ell}, \sigma_\ell^2, \tau_\ell$ for the estimators obtained under the constraint $\theta_2 = \ell$. With this notation,

$$
\mathcal{S}_{\mathrm{L}} = -2\log \frac{L_n(Y_1, \ldots, Y_n; \widehat{\theta}_1, \widehat{\theta}_2, \widehat{\sigma}^2, \widehat{\tau})}{L_n(Y_1, \ldots, Y_n; \widehat{\theta}_{1\,\ell}, \ell, \widehat{\sigma}_\ell^2, \widehat{\tau}_\ell)} ,
$$

and the decision rule is to reject H if $\mathcal{S}_{\mathrm{L}} > C$, with C given by

$$
\Pr\{Z_1 > C\} = \alpha,
$$

Z_1 being distributed as a χ^2 with 1 degree of freedom. We now deduce a confidence region from this test procedure. Obviously, $\mathcal{S}_{\mathrm{L}}$ depends on ℓ and we denote by $\mathcal{S}_{\mathrm{L}}(\ell)$ the value taken by $\mathcal{S}_{\mathrm{L}}$ for a given ℓ. We can define as a confidence region for θ_2 the set of all of the real ℓ such that $\mathcal{S}_{\mathrm{L}}(\ell) \leq C$. In other words, we consider as a confidence region for θ_2 with asymptotic confidence level $1 - \alpha$ the set of all of the real values ℓ such that we do not reject the hypothesis "H: $\{\theta_2 = \ell\}$" using a likelihood ratio test with asymptotic level α.

Remarks

1. The confidence region $\{\ell : \mathcal{S}_{\mathrm{L}}(\ell) \leq C\}$ cannot be translated into a more explicit form. We use some numerical computations to obtain its endpoints. Note that it is not necessarily an interval, but in practical studies it turns out to be one generally.

2. This type of confidence region based on the log-likelihood ratio can easily be generalized to any subset of the parameter vector ϕ, or to any linear function of the type $\Lambda\phi = L_0$, with Λ an $r \times p + q + 1$ matrix of rank r, $r < p + q + 1$ and L_0 an r-dimensional vector.

3.5 Applications

3.5.1 Growth of winter wheat tillers

Model. The regression function is

$$
f(x, \theta) = \theta_1 \exp(\theta_2 x),
$$

and the variance function is $\mathrm{Var}(\varepsilon_i) = \sigma^2 f(x_i, \theta)$.

Method. The parameters are estimated by maximizing the log-likelihood, $V(\theta, \sigma^2, \tau)$; see equation (3.8).

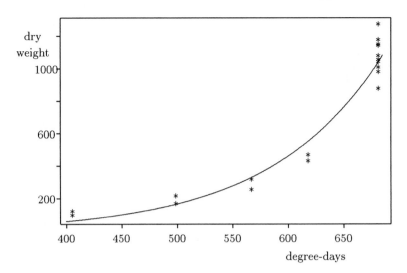

FIGURE 3.4. Weight of tillers example: graph of observed and adjusted curve

Results.

Parameters	Estimated values	Asymptotic covariance matrix	
θ_1	1.14	0.24	
θ_2	0.01	-0.0003	4 10^{-7}
σ^2	13.32		

The adjusted response curve, $f(x,\widehat{\theta})$, is shown in figure 3.4. A quick inspection of this graph shows that the regression model is not very well chosen. It would be better to have a regression function with a more gradual increase for low values of x. This can be achieved by elevating $\theta_2 x$ to some power, that is, by choosing

$$f(x,\theta) = \theta_1 \exp\left[(\theta_2 x)^{\theta_3}\right] , \qquad (3.10)$$

the variance function being as before, $\text{Var}(\varepsilon_i) = \sigma^2 f(x_i, \theta)$.

Results.

Parameters	Estimated values	Asymptotic covariance matrix		
θ_1	79.13	702.2		
θ_2	0.0019	-0.0004	2 10^{-8}	
θ_3	4.05	23.07	1 10^{-8}	0.85
σ^2	7.30			

Likelihood ratio test of "H: $\theta_3 = 1$" against "A: $\theta_3 \neq 1$" using results of section 3.4.2. Figure 3.5 suggests that the regression model given by equation (3.10) fits the tiller data better.

To validate this observation, we perform a likelihood ratio test. The hypothesis to be tested is $\{\theta_3 = 1\}$. In this case, the hypothesis is described

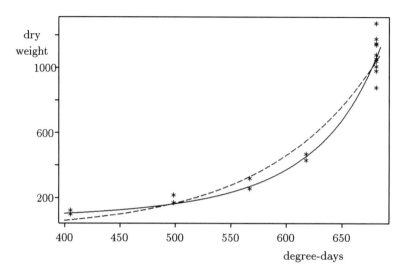

FIGURE 3.5. Weight of tillers example: graph of observed and adjusted curves for both the hypothesis H (dashed line) and the alternative A (solid line)

by model (3.1), and the alternative is described by model (3.10). So $\widehat{\theta}_{1\,\mathrm{H}} = 1.14$, $\widehat{\theta}_{2\,\mathrm{H}} = 0.01$ and $\widehat{\theta}_{3\,\mathrm{H}} = 1$; whereas $\widehat{\theta}_1 = 79.13$, $\widehat{\theta}_2 = 0.0019$ and $\widehat{\theta}_3 = 4.05$. The statistic S_L is equal to 10.4. If Z_1 is distributed as a χ^2 with 1 degree of freedom, the critical value C such that $\Pr\{Z_1 \le C\} = 0.95$ is 3.84. Thus, using the likelihood ratio test with an asymptotic 5 % level, we reject the hypothesis H.

Calculation of a confidence interval for θ_3 using the asymptotic distribution of $\widehat{\theta}_3$. See section 3.4.4.

$\widehat{\theta}_3$	$\widehat{S}$	$\nu_{0.975}$	$\widehat{I}_{\mathcal{N}}$
4.05	0.924	1.96	[2.239 , 5.861]

Calculation of a confidence interval for θ_3 using the log-likelihood ratio. See section 3.4.4.

As mentioned above, the computation of a confidence interval based on the log-likelihood ratio requires some extra calculations after we have performed the estimation. Although these calculations are implemented in **S-PLUS**, (see section 3.6), describing the necessary calculations as follows helps to understand the method better:

1. Choose a *reasonable* set of possible values for θ_3. Here, $\widehat{\theta}_3$ is equal to 4.05, and one estimate of its standard error $\widehat{S}$ is equal to 0.924. We take 39 equispaced values $t_1, t_2, \ldots, t_{39}$ with $t_1 = 1$ and $t_{39} = 7.10$; the central value t_{20} is just $\widehat{\theta}_3$. Note that the choice of this series is quite arbitrary. We decided to explore the behavior of the log-likelihood ratio for values of θ_3 roughly between $\widehat{\theta}_3 - 3\widehat{S}$ and $\widehat{\theta}_3 + 3\widehat{S}$.

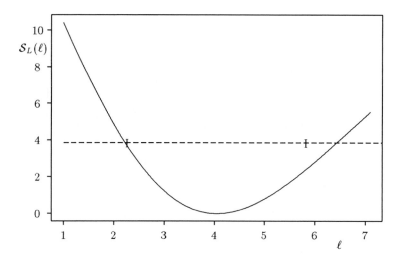

FIGURE 3.6. Confidence intervals: values of the log-likelihood ratio when the parameter θ_3 varies $(\theta_3 = \ell)$

2. For each value t_i, for $i = 1, \ldots, 39$, fix $\theta_3 = t_i$ and estimate the other parameters under this constraint. Then compute the log-likelihood ratio $S_L(t_i)$ (obviously, $S_L(t_{20}) = 0$).

3. Draw a picture of $S_L(\ell)$ versus ℓ. This is a useful step to check the regularity of the log-likelihood and to provide a comprehensive look at the results of the preceding step. On the same graph, draw the horizontal line corresponding to the equation $y = C$. The abscissa of the intersection points between this line and the graph of the log-likelihood are the endpoints of the confidence region based on the log-likelihood ratio.

4. Determine the endpoints of the confidence region by linear interpolation. Here the picture (see figure 3.6) clearly shows that the region is an interval. Let us consider the set of values $S_L(t_i)$, for $i = 1, \ldots, 39$ (see table 3.3). For a confidence interval with asymptotic level 0.95, choose C such that $\Pr\{Z_1 > C\} = 0.05$, for Z_1 a χ^2 with 1 degree of freedom: $C = 3.8415$. From table 3.3, we deduce that the left endpoint of the log-likelihood ratio based confidence interval is between $t_8 = 2.1237$ and $t_9 = 2.2842$, because $S_L(t_8) = 4.3269$ and $S_L(t_9) = 3.6335$. A precise enough value of the lower bound $\widehat{\theta}_{3\,\text{inf}}$ of the confidence interval is given by linear interpolation and is 2.236. Similarly, the right endpoint, or upper bound, $\widehat{\theta}_{3\,\text{sup}}$ of the confidence interval is between t_{34} and t_{35} and is 6.428.

TABLE 3.3. Log-likelihood values

t	$S_{\mathrm{L}}(t)$	t	$S_{\mathrm{L}}(t)$	t	$S_{\mathrm{L}}(t)$
1.0000	10.4256	3.0868	1.0491	5.1737	1.0856
1.1605	9.4496	3.2474	0.7203	5.3342	1.3781
1.3211	8.5016	3.4079	0.4550	5.4947	1.6940
1.4816	7.5856	3.5684	0.2523	5.6553	2.0302
1.6421	6.7060	3.7289	0.1104	5.8158	2.3835
1.8026	5.8669	3.8895	0.0272	5.9763	2.7513
1.9632	5.0726	4.0500	0.0000	6.1368	3.1308
2.1237	4.3269	4.2105	0.0256	6.2974	3.5198
2.2842	3.6335	4.3711	0.1006	6.4579	3.9162
2.4447	2.9957	4.5316	0.2214	6.6184	4.3180
2.6053	2.4164	4.6921	0.3841	6.7789	4.7236
2.7658	1.8978	4.8526	0.5849	6.9395	5.1315
2.9263	1.4417	5.0132	0.8200	7.1000	5.5403

Let us summarize the results of this procedure in the following table:

$\widehat{\theta}_3$	$\widehat{S}$	$\chi^1_{1,0.95}$	$\widehat{I}_{\mathcal{S}}$
4.05	0.924	3.8415	[2.236 , 6.428]

Comparing $\widehat{I}_{\mathcal{S}}$ with $\widehat{I}_{\mathcal{N}}$, whose endpoints appear as vertical bars on the horizontal dashed line of figure 3.6, we observe that $\widehat{I}_{\mathcal{S}}$ is larger and non-symmetric around $\widehat{\theta}_3$. Obviously, the likelihood takes into account the regression function as it is, whereas using the Wald statistic we consider that the model is almost a linear regression. Both confidence intervals are based on asymptotic approximations, but for $\widehat{I}_{\mathcal{N}}$ this is even more drastic.

3.5.2 Solubility of peptides in trichloacetic acid solutions

Model. The regression function is

$$f(x,\theta) = \frac{100}{1 + \exp\left(\theta_2\left(x - \theta_1\right)\right)},$$

and the variance function is $\mathrm{Var}(\varepsilon_i) = \sigma^2\left(1 + \tau f(x_i, \theta)\left(100 - f(x_i,\theta)\right)\right)$.

Method. The parameters are estimated by maximizing the log-likelihood, $V(\theta, \sigma^2, \tau)$; see equation (3.8).

Results.

Parameters	Estimated values	Asymptotic covariance matrix		
θ_1	43.93	0.3946		
θ_2	0.4332	$-1.5\ 10^{-6}$	0.0031	
τ	0.0245	$-6.1\ 10^{-5}$	0.0004	$8.1\ 10^{-5}$
σ^2	26.79			

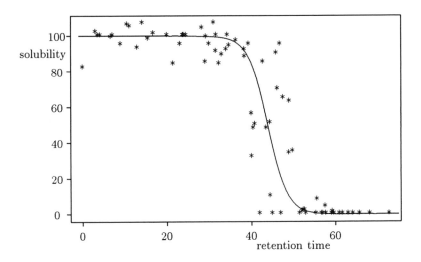

FIGURE 3.7. Peptides example: observed and adjusted solubilities of 75 peptides versus their RP-HPLC retention time

The adjusted response curve, $f(x, \widehat{\theta})$, is shown in figure 3.7.

Likelihood ratio test of the existence of a constant term in the variance model. In order to verify that a constant term in the variance is necessary to correctly describe the variability of the data, we perform a test of the hypothesis H: $\{\mathrm{Var}(\varepsilon_i) = \sigma^2 f(x_i, \theta)\,(100 - f(x_i, \theta))\}$. In order to make clear that this is a hypothesis of the type $\Lambda\phi = L_0$, let us rewrite the variance model under A as $\{\mathrm{Var}(\varepsilon_i) = \sigma_1^2 + \sigma_2^2 f(x_i, \theta)\,(100 - f(x_i, \theta))\}$. In this notation, we express the hypothesis H as $\{\sigma_1^2 = 0\}$. Between this alternative modeling of the variance and the original one given by equation (3.3), there is only a simple one-to-one transformation: $\sigma_1^2 = \sigma^2$, $\sigma_2^2 = \sigma^2 \tau$. The values of the estimates under this assumption are $\widehat{\theta}_{1\,\mathrm{H}} = 41.70$ and $\widehat{\theta}_{2\,\mathrm{H}} = 0.1305$. The statistic S_{L} is equal to 70.14. Because it is a highly unlikely value for a χ^2 random variable with 1 degree of freedom, the hypothesis H must be rejected.

Figure 3.8 illustrates the observed solubilities, drawn together with the adjusted one under the hypothesis H (dashed line) and model (3.2) (solid line). The second line is obviously a better fit.

A discerning reader could have foreseen the result of this test. Indeed, assuming that the hypothesis H is fulfilled is equivalent to assuming that the extreme measures (corresponding to $f(x, \theta) = 0$ or $f(x, \theta) = 100$) are performed with perfect accuracy. Thus the estimation process gives these extreme measures a very high weight, whereas it barely takes the intermediate data into account. The dashed curve in figure 3.8, which corresponds to the peptide data fit under the hypothesis H, illustrates this fact.

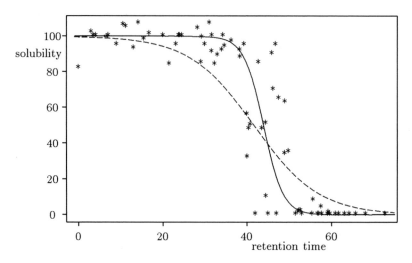

FIGURE 3.8. Peptides example: observed and adjusted solubilities of 75 peptides versus their RP-HPLC retention time for both the hypothesis H (dashed line) and the alternative A (solid line)

Calculation of a confidence interval for θ_2 using the log-likelihood ratio. Using the same type of procedure that we used to compute a confidence interval for θ_3 in section 3.5.1, we obtain the following result:

$\widehat{\theta}_2$	$\widehat{S}$	$\chi^2_{1,0.95}$	$\widehat{I}_{\mathcal{S}}$
0.433	0.0557	3.8415	[0.309 , 0.670]

Calculation of a confidence interval for θ_2 using the Wald statistic, based on the maximum likelihood estimator. Standard calculations lead to the following result :

$\widehat{\theta}_2$	$\widehat{S}$	$\nu_{0.975}$	$\widehat{I}_{\mathcal{N}}$
0.433	0.0557	1.96	[0.324, 0.542]

Calculation of a confidence interval for θ_2 using the Wald statistic, based on the three-step alternate method. We introduce, for the sake of comparison, the three-step alternate method for estimating the parameters, using an estimator other than the maximum likelihood estimator. We choose ordinary least squares to estimate θ during the first step, modified least squares to estimate β during the second step, and modified least squares to estimate θ during the third step. We obtain the following results.

Parameters	Estimated values	Asymptotic covariance matrix	
θ_1	43.81	0.7427	
θ_2	0.233	5.810^{-4}	0.00101
τ	0.0084		
σ^2	27.56		

Both estimated regression functions are shown in figure 3.9.

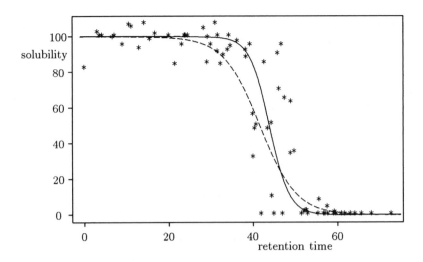

FIGURE 3.9. Peptides example: observed and adjusted solubilities of 75 peptides versus their RP-HPLC retention time for both the three-step alternate method (dashed line) and the maximum likelihood method (solid line)

We can also compute the confidence interval for θ_2 based on the results of this estimation procedure:

$\widehat{\theta}_2$	$\widehat{S}$	$\nu_{0.975}$	$\widehat{I'_{\mathcal{N}}}$
0.233	0.031	1.96	[0.171, 0.295]

Actually, in this example, the results of both estimation methods disagree to a large extent. As far as we trust our modeling of the phenomenon, the three-step alternate method, using modified least squares methods at steps 2 and 3, does not fit the data sufficiently well, as we can see from an examination of figure 3.9 and also by comparing the value of the log-likelihood obtained by both methods: the log-likelihood computed with the values of the three-step alternate estimators is -281.8, whereas the maximum log-likelihood is -288.7. This strongly supports our advice: use the maximum likelihood method as often as you can!

3.6   Using **nls2**

This section reproduces the commands and files used in this chapter to treat the examples by using **nls2**.

Growth of winter wheat tiller example

The wheat tiller example is a new example, just introduced in this chapter. We first create the *data-frame* to store the experimental data. Then, we estimate the parameters with two models, and, finally, we calculate confidence intervals to compare them.

Creating the data

The experimental data (see table 3.1, page 62) are stored in a *data-frame* called `tiller`:

```
> DegreeDays <-c(
    405.65, 405.65, 498.75, 498.75, 567.25, 567.25, 618.30,
    618.30, 681.45, 681.45, 681.45, 681.45, 681.45, 681.45,
    681.45, 681.45, 681.45, 681.45)
> DryWeight <-c(
    113.386,  90.500, 161.600, 207.650, 309.514, 246.743,
    460.686, 422.936, 1047.000, 972.383, 1072.022, 1034.000,
    1169.767, 1141.883, 999.633, 1266.290, 868.662, 1133.287)
  tiller <- data.frame( DegreeDays,  DryWeight)
```

Plot of the observed dry weights on a degree days time scale.

We plot the observed response values versus the degree-days scale by using the graphical function `plot` of **S-PLUS**:

```
> plot(DegreeDays,DryWeight,xlab="degree days",ylab="dry weight",
        main="Growth of winter wheat tillers example",
        sub="Observed response")
```

(See figure 3.1, page 62).

Plot of the empirical variance versus mean

S-PLUS functions are used to calculate empirical intrareplications variances and means. We plot the empirical variances versus the empirical means and the same plot after the logarithm transformation:

```
> tiller.mean<-as.vector(
        unlist(lapply(split(DryWeight,DegreeDays),mean)))
> tiller.var<-as.vector(
        unlist(lapply(split(DryWeight,DegreeDays),var)))
> plot(tiller.mean, tiller.var, xlab="mean", ylab="var",
        main="Growth of winter wheat tillers example",
        sub="Empirical response variance versus response mean")
> plot(log(tiller.mean), log(tiller.var),
        xlab="log(mean)", ylab="log(var)",
        main="Growth of winter wheat tillers example",
        sub="Log of empirical response variance\
  versus log of response mean")
```

(See figure 3.2, page 63).

Parameters estimation under hypothesis H

We describe the model of hypothesis H (the simple exponential function defined in section 3.5.1, page 72) in a file called `tiller.exp.m2`:

```
% Model simple exponential
%        with response variances proportional
%        to response expectations
resp DryWeight;
varind DegreeDays;
var v;
parresp a,b;
subroutine;
begin
DryWeight= a*exp(b*DegreeDays);
v = DryWeight;
end
```

To find reasonable starting values for the parameters, we make a linear regression of log(dry weight) on degree-days by using the function lm in **S-PLUS**. Then we call **nls2**. We display the estimated values of the parameters $\widehat{\sigma}^2$ and the estimated asymptotic covariance matrix:

```
> tiller.lm <- lm(log(DryWeight) ~ DegreeDays)
> tiller.nl2 <- nls2(tiller,  "tiller.exp.m2",
    c(exp(tiller.lm$coefficients[1]),tiller.lm$coefficients[2]))
>     # Print the results:
> cat( "Estimated values of the parameters:\n ")
> print(tiller.nl2$theta)
> cat( "\nEstimated value of sigma2:\n ")
> print( tiller.nl2$sigma2)
> cat( "\nEstimated asymptotic covariance matrix: \n ")
> print(tiller.nl2$as.var); cat("\n\n")
```

We plot the observed and fitted response values versus the degree-days scale:

```
> plot(DegreeDays,DryWeight,xlab="degree days",ylab="dry weight",
        main="Growth of winter wheat tillers example",
        sub="Observed response")
> X<-seq(400,685,length=100)
> Y<-tiller.nl2$theta[1]*exp(tiller.nl2$theta[2]*X)
> lines(X,Y)
```

(See figure 3.4, page 73).

Parameters estimation under hypothesis A

The second model to test is the double exponential function defined by equation (3.10), page 73. We describe it in a file called tiller.exp.m7:

```
% Model double exponential
%        with response variances proportional
%        to response expectations
resp DryWeight;
varind DegreeDays;
var v;
parresp a,b,g;
subroutine;
```

```
begin
DryWeight= a*exp(exp(g*log(DegreeDays*b)));
v = DryWeight;
end
```

The function nls2 is called to carry out the estimation. The values estimated with the first model are assigned to the starting values of the parameters. Option max.iters increases the default maximal number of iterations, so we can be sure that convergence will be reached:

```
> tiller.nl7 <- nls2(tiller, "tiller.exp.m7",
    list(theta.start = c(tiller.nl2$theta, 1), max.iters=500))
>    # Print the results:
> cat( "Estimated values of the parameters:\n ")
> print(tiller.nl7$theta)
> cat( "\nEstimated value of sigma2:\n ")
> print( tiller.nl7$sigma2)
> cat( "\nEstimated asymptotic covariance matrix: \n ")
> print(tiller.nl7$as.var); cat("\n\n")
```

Plots of the fitted curves

The observed responses, the responses fitted under hypothesis H (joined by a dashed line), and the responses fitted under hypothesis A (joined by a solid line) are plotted versus the degree-days scale:

```
> plot(DegreeDays,DryWeight,xlab="DegreeDays",ylab="dry weight",
    main="Growth of winter wheat tillers example",
    sub="Observed response")
> X<-seq(400,685,length=100)
> Y<-tiller.nl7$theta[1] *
        exp((tiller.nl7$theta[2]*X)**tiller.nl7$theta[3])
> lines(X,Y)
> Z<-tiller.nl2$theta[1]*exp(tiller.nl2$theta[2]*X)
> par (lty=2)
> lines (X,Z)
> par (lty=1)
```

(See figure 3.5, page 74).

Likelihood ratio test of "H: $\theta_3 = 1$" against "A: $\theta_3 \neq 1$"

To be sure that the second model fits the tiller data better, we test the hypothesis "H: $\theta_3 = 1$" against the alternative "A: $\theta_3 \neq 1$" using a likelihood ratio test. We calculate and display the statistic S_L (see equation (3.9), page 70) and the 0.95 quantile of a χ^2 with 1 degree of freedom with which it should be compared:

```
> cat( "Sl:",
    dim(tiller)[1] * (tiller.nl2$loglik - tiller.nl7$loglik),
    "X2(0.95,1):",  qchisq(0.95,1), "\n\n")
```

Confidence interval for θ_3 using the asymptotic distribution of $\widehat{\theta}_3$

We calculate the confidence interval $\widehat{I}_\mathcal{N}$ (see equation (2.2), page 32, and section 3.4.4, page 71) by using the function `confidence`. We display the values of $\widehat{\theta}_3$, $\widehat{S}$, $\nu_{0.975}$ and $\widehat{I}_\mathcal{N}$:

```
> tiller.In <- confidence(tiller.nl7)
>   # Print the results:
> cat("Estimated value of theta_3:", tiller.nl7$theta[3],"\n" )
> cat("Estimated value of S:", sqrt(tiller.nl7$as.var[3, 3]),"\n" )
> cat("nu_(0.975):", qnorm(0.975),"\n" )
> cat("Estimated value of In:",tiller.In$normal.conf.int[3,],"\n" )
```

Confidence interval for θ_3 using the log-likelihood ratio

To calculate the confidence interval based on the log-likelihood ratio (see section 3.4.4, page 71, and $\widehat{I}_S$ section 3.5.1, page 74), we use the function `conflike`. We display the values of $\widehat{\theta}_3$, $\widehat{S}$, $\chi^1_{1,0.95}$ and $\widehat{I}_S$:

```
> cat("Estimated value of theta_3:", tiller.nl7$theta[3],"\n" )
> cat("Estimated value of S:", sqrt(tiller.nl7$as.var[3, 3]),"\n" )
> cat("X2(0.95,1):", qchisq(0.95, 1),"\n" )
> cat("Estimated value of Is:",
    conflike(tiller.nl7,parameter=3)$like.conf.int,"\n" )
```

(Results for this example are given in section 3.5.1, page 72.)

The function `conflike` also allows us to carry out all of the steps of the calculation shown page 74.

Solubility of peptides example

The solubility of peptides example is a new example, just introduced in this chapter. We first create a *data-frame* to store the experimental data. Then we estimate the parameters with several models, and, finally, we calculate confidence intervals to compare them.

Creating the data

The experimental data (see table 3.2, page 64) are stored in a *data-frame* called **pept.d**:

```
> RetTime <-c(
  3.6,   3.6,   6.7, 11.2, 20.0, 21.5,
  24.5, 29.2, 38.4, 40.4, 49.8, 59.6,
  64.2, 68.1, 72.7, 40.8, 52.7, 57.7,
  59.1, 61.7, 65.7, 28.3, 33.0, 40.0,
  44.5, 47.0, 63.1, 31.6, 40.1, 52.4,
  57.5, 23.8, 29.1, 30.0, 31.7, 23.1,
  42.7, 46.8, 16.8, 24.0, 31.1, 32.3,
  34.0, 34.7, 36.3, 39.3, 42.0, 44.4,
  45.0, 46.2, 49.0, 52.1, 55.4, 56.7,
  53.0,  0.0,  3.1,  4.2,  7.1, 9.0, 10.6,
```

84   3. Variance estimation

```
 13.0, 14.2, 15.5, 38.3, 43.5, 45.9,
 47.5, 48.9, 55.6, 57.0, 59.2, 60.9, 34.3, 51.5)
> solubility  <-c(
 100, 100, 99, 105, 100, 84, 100, 99, 88, 48, 35, 0,
  0, 0, 0, 50, 2, 0, 0, 0, 0, 104, 89, 56, 10, 0, 0,
 91, 32, 2, 4, 100, 85, 95, 100, 95, 85, 95, 101, 100,
 107, 84, 92, 94, 97, 95, 0, 51, 0, 70, 63, 1, 0, 0,
 0, 82, 102, 100, 100, 95, 106, 93, 107, 98, 92, 48,
 90, 65, 34, 8, 0, 1, 0, 100, 0)
> pept.d <- data.frame(RetTime , solubility)
```

Plot of the observed percentage of solubility versus the retention time

We plot the observed responses versus the retention time by using the
graphical function pldnls2:

```
> pldnls2(pept.d, response.name="solubility", X.name="RetTime",
          title="Solubility of peptides example",
          sub="observed response")
```

(See figure 3.3, page 65).

Find initial values for the regression parameters

To find reasonable starting values for the parameters, we make a first estimation using ordinary least squares estimators and assume constant variance. This model is described in a file called pept.m1:

```
% model pept.m1
resp solubility;
varind RetTime;
aux a1;
parresp  ed50, sl ;
subroutine ;
begin
a1 = 1 + exp (sl*(RetTime-ed50)) ;
solubility = 100. /a1 ;
end
```

We apply function nls2 to estimate the parameters:

```
> pept.nl1<-nls2(pept.d, "pept.m1",
    list(theta.start=c(1,1), max.iters=100))
```

Estimation under hypothesis A

To see whether a constant term in the variance is necessary to describe
correctly the variability of the data, we first try a variance model with a
constant term. We describe it in a file called pept.m3:

```
% model pept.m3
resp solubility;
varind RetTime;
var v;
```

```
aux a1;
parresp  ed50, sl ;
parvar  h;
subroutine ;
begin
a1 = 1 + exp (sl*(RetTime-ed50)) ;
solubility = 100. /a1 ;
v= 1 + h*solubility*(100.-solubility)   ;
end
```

We call **nls2** and display the results:

```
>  pept.c3 <- list(theta.start=pept.nl1$theta,
                   beta.start=0)
>  pept.nl3<-nls2(pept.d, "pept.m3", stat.ctx= pept.c3)
>      # Print the results
> cat("Summary of pept.nl3:\n")
> summary( pept.nl3)
```

We plot the observed and fitted responses versus the retention time by using the graphical function **plfit**. The option **wanted** specifies the requested plot: the observed and fitted values are plotted against the independent variable:

```
> plfit(pept.nl3, wanted=list(X.OF=T),
        title="Solubility of peptides example",
        sub="Observed and fitted response")
```

(See figure 3.7, page 77).

Estimation under hypothesis H

Now we try a variance model without a constant term. This model is described in a file called **pept.m2**:

```
% model pept.m2
resp solubility;
varind RetTime;
var v;
aux a1;
parresp  ed50, sl ;
subroutine ;
begin
a1 = 1 + exp (sl*(RetTime-ed50)) ;
solubility = 100. /a1 ;
v= solubility*(100.-solubility) ;
end
```

The parameters of this model are estimated by using the function **nls2**:

```
> pept.nl2<-nls2(pept.d, "pept.m2",pept.nl4$theta)
```

Likelihood ratio test of the existence of a constant term in the variance model

We calculate and display the test statistic $\mathcal{S}_L$ (see equation (3.9), page 70) and the 0.95 quantile of a χ^2 with 1 degree of freedom, with which $\mathcal{S}_L$ should be compared:

```
> cat( "Sl:",
    75 * (pept.nl2$loglik - pept.nl3$loglik),
    "X2(0.95,1):",  qchisq(0.95,1), "\n\n")
```

Plots of the fitted curves

The observed responses, the responses fitted under hypothesis H (joined by a dashed line), and the responses fitted under hypothesis A (joined by a solid line) are plotted versus the retention time:

```
> plot(RetTime,solubility,xlab="retention time",ylab="solubility",
main="Solubility of peptides example",
        sub="Observed and fitted response")
> X<-seq(-1,75,length=100)
> Y<-100/(1+exp(pept.nl3$theta[2]*(X-pept.nl3$theta[1])))
> lines(X,Y)
> Z<-100/(1+exp(pept.nl2$theta[2]*(X-pept.nl2$theta[1])))
> par (lty=2)
> lines (X,Z)
> par (lty=1)
```

(See figure 3.8, page 78).

Confidence interval for θ_2 using the log-likelihood ratio

We calculate a confidence interval based on the log-likelihood ratio (see section 3.4.4, page 71, and $\widehat{I}_S$ section 3.5.1, page 74) by using the function conflike. We display the values of $\widehat{\theta}_2$, $\widehat{S}$, $\chi^1_{1,0.95}$ and $\widehat{I}_S$:

```
> pept.conflike _ conflike(pept.nl3, parameter = 2)
>    # Print the results
> cat("Estimated value of theta_2:", pept.nl3$theta[2],"\n" )
> cat("Estimated value of S:",  sqrt(pept.nl3$as.var[2, 2]),"\n" )
> cat("X2(0.95,1):", qchisq(0.95, 1),"\n" )
> cat("Estimated value of Is:", pept.conflike$like.conf.int,"\n" )
```

Confidence interval for θ_2 using the Wald statistic, based on the maximum likelihood estimator

To calculate a confidence interval for θ_2 using the Wald statistic (see section 3.4.3, page 70) we apply the function confidence. We display the values of $\widehat{\theta}_2$, $\widehat{S}$, $\nu_{0.975}$ and $\widehat{I}_N$:

```
> pept.conf_confidence(pept.nl3)
>    # Print the results:
> cat("Estimated value of theta_2:",pept.conf$psi[2],"\n" )
```

```
> cat("Estimated value of S:", pept.conf$std.error[2],"\n" )
> cat("nu_(0.975):", qnorm(0.975),"\n" )
> cat("Estimated value of In:",pept.conf$normal.conf.int[2,], "\n" )
```

Confidence interval for θ_2 using the Wald statistic, based on the three-step alternate method

The alternate method is included in the function **nls2**: option **method** introduces the three requested methods. They are as follows; **OLST**, "Ordinary Least Square," to estimate the parameters θ; **MLSB**, "Modified Least Square," to estimate the parameters β; and **MLST**, "Modified Least Square," to reestimate the parameters θ.

We display the values of the parameters and the asymptotic covariance matrix estimated at the last step:

```
> pept.nl8<-nls2(pept.d, "pept.m3", pept.c3,
    method=c("OLST","MLSB","MLST"))
>    # Print the results
> cat("Estimated value of the parameters theta:\n")
> print(pept.nl8$step3$theta)
> cat("Estimated value of the parameters beta:\n",
    pept.nl8$step3$beta[1], pept.nl8$step3$sigma2)
> cat("\nEstimated value of the asymptotic covariance matrix:\n")
> print(pept.nl8$step3$as.var[1:2, 1:2])
```

Plots of the results

The observed responses, the responses fitted by the three-step alternate method (joined by a dashed line), and the responses fitted by the maximum likelihood method (joined by a solid line) are plotted versus the retention time:

```
> plot(RetTime,solubility,
        xlab="retention time",ylab="solubility",
        main="Solubility of peptides example",
        sub="Observed and fitted response")
> X<-seq(-1,75,length=100)
> Y<-100/(1+exp(pept.nl3$theta[2]*(X-pept.nl3$theta[1])))
> Z<-100/(1+exp(pept.nl8$step3$theta[2]*(X-pept.nl2$theta[1])))
> lines(X,Y)
> par (lty=2)
> lines (X,Z)
> par (lty=1)
```

(See figure 3.9, page 79).

Confidence interval for θ_2 based on results of the three-step method

Using the function **confidence**, we compute the confidence interval based on the Wald statistic from the results of estimation by the alternate method. We display the values of $\hat{\theta}_2$, $\hat{S}$, $\nu_{0.975}$ and $\hat{I}'_N$:

```
> pept.In <- confidence(pept.nl8)
```

```
>       # Print the results
> cat("Estimated value of theta_2:",coef(pept.nl8)$theta[2],"\n" )
> cat("Estimated value of S:", coef(pept.nl8)$std.error[2],"\n" )
> cat("nu_(0.975):", qnorm(0.975),"\n" )
> cat("Estimated value of In:",pept.In$normal.conf.int[2,],"\n" )
```

We display the log-likelihood values calculated by both methods to compare them:

```
> cat("Loglikelihood values:",
    -(pept.nl3$loglik*75/2), -(pept.nl8$step3$loglik*75/2),"\n" )
```

(Results for this example are given in section 3.5.2, page 76.)

4

Diagnostics of model misspecification

In section 1.2, we denoted the true regression relationship between Y and x by $\mu(x)$, and we noted that $\mu(x)$ is generally unknown, although in practice it is approximated by a parametric function $f(x, \theta)$. Because we use this f to make some statistical inferences, however, we must be able to determine whether or not our choice of f is accurate.

We can detect model misspecification by examining the data together with the results of the estimation. It is an iterative procedure. The goal is to check if the assumptions on which the analysis is based are fairly accurate and to detect and correct any existing model misspecification. For this purpose we present two types of tools in this chapter: graphics and tests, applying each of these, respectively, to some of our previously cited examples.

4.1 Problem formulation

Let us recall the assumptions: we observe (Y_{ij}, x_i), $i = 1, \ldots k$, $j = 1, n_i$. We consider the parametric functions f and g such that

$$
\begin{aligned}
Y_{ij} &= f(x_i, \theta) + \varepsilon_{ij} \\
\mathrm{Var}\varepsilon_{ij} &= g(x_i, \sigma^2, \theta, \tau).
\end{aligned}
$$

We assume that the errors ε_{ij} are independent for all (i, j).

Then we estimate the parameters. Before we describe how to detect misspecifications, let us give the list of assumptions that these procedures need to check.

- The regression function f is correct.

  The true regression function $\mu(x)$ should verify $\mu(x) = f(x, \theta)$ for one value of θ in Θ, and for all values of x. If this condition exists, it implies that the errors are centered variables: $E\varepsilon_{ij} = 0$. If not, $EY_{ij} = \mu(x_i)$ and $E\varepsilon_{ij} = \mu(x_i) - f(x_i, \theta)$, which is different from zero for at least some values of x_i, $i = 1, \ldots n$. In this latter case, the *classical asymptotic results* stated in section 2.3 no longer are valid.

- The variance function is correct.

  The most frequent mistake made in modeling the variance function is to assume that the variance of errors is constant, $\mathrm{Var}\varepsilon_{ij} = \sigma^2$, when, in actuality, heteroscedasticity exists, $\mathrm{Var}\varepsilon_{ij} = \sigma_i^2$. In that case, when the number of observations tends to infinity, the variance of $V_{\widehat{\theta}}^{-1/2}(\widehat{\theta} - \theta)$ is different from the identity matrix.

- The observations are independent.

  Suppose that we are interested in the growth of some characteristic as a function of time. We observe the response Y_i at time t_i for one individual, for $i = 1, \ldots n$. Generally, the relationship between the growth and the time, say $f(t, \theta)$, is a smooth function of t. If, however, the t_i are such that the $t_{i+1} - t_i$ are small with respect to the variations of the growth, correlations exist between the variables Y_i. The observations thus are not independent, and the results of section 2.3 cannot be applied, as we will see in section 4.2.5.

4.2 Diagnostics of model misspecifications with graphics

Graphic of fitted and observed values

If the Y_i verify $EY_i = f(x_i, \theta)$, then the fitted values $f(x_i, \widehat{\theta})$ estimate the expectation of the Y_i. To detect a bad choice of the regression function, we use either a plot of fitted $f(x_i, \widehat{\theta})$ and observed Y_i values versus x_i, or a plot of fitted versus observed values. These graphics tools offer a simple way to look simultaneously at data and fitted values.

Graphics of residuals and standardized residuals

If the choice of the regression function is right, then the

$$\varepsilon_{ij} = Y_{ij} - f(x_i, \theta), \text{ for } i = 1, \ldots k, \ j = 1, \ldots n_i$$

are centered, independent variables; and if $\mathrm{Var}\varepsilon_{ij} = \sigma_i^2$, the standardized errors

$$e_{ij} = \frac{Y_{ij} - f(x_i, \theta)}{\sigma_i}, \text{ for } i = 1, \ldots k, \ j = 1, \ldots n_i$$

are centered, independent variables with variance equal to 1. In the same way, if the variance function is $\sigma_i^2 = g(x_i, \sigma^2, \theta, \tau)$, then the

$$e_{ij} = \frac{Y_{ij} - f(x_i, \theta)}{\sqrt{g(x_i, \sigma^2, \theta, \tau)}}, \text{ for } i = 1, \ldots k, \quad j = 1, \ldots n_i$$

are centered, independent variables with variance equal to 1. A natural idea is to estimate the errors and the standardized errors and then study their behavior.

The residuals

$$\widehat{\varepsilon}_{ij} = Y_{ij} - f(x_i, \widehat{\theta})$$

and the standardized residuals

$$\widehat{e}_{ij} = \frac{Y_{ij} - f(x_i, \widehat{\theta})}{\widehat{\sigma}_i}$$

are natural estimations of the errors and standardized errors. Model misspecification usually is detected by examining the graphics of residuals. To detect a bad choice of the regression or variance functions, we use the graphics of $\widehat{\varepsilon}_{ij}$, $\widehat{e}_{ij}$, $|\widehat{\varepsilon}_{ij}|$, $|\widehat{e}_{ij}|$, $\widehat{\varepsilon}_{ij}^2$, $\widehat{e}_{ij}^2$ versus x_i, $f(x_i, \widehat{\theta})$, or $g(x_i, \widehat{\sigma}^2, \widehat{\theta}, \widehat{\tau})$. If $n_i = 1$, the graphic of $\widehat{\varepsilon}_i$ versus $\widehat{\varepsilon}_{i-1}$ is convenient to detect some correlation between errors.

4.2.1 Pasture regrowth example: estimation using a concave-shaped curve and plot for diagnostics

Let us again consider the pasture regrowth example using a concave-shaped curve for the function f in place of function (1.2).

Model. The regression function is

$$f(x, \theta) = \frac{\theta_2 + \theta_1 x}{\theta_3 + x}, \tag{4.1}$$

and the variances are homogeneous: $\mathrm{Var}(\varepsilon_i) = \sigma^2$.

Method. The parameters are estimated by minimizing the sum of squares, $C(\theta)$; see equation (1.10).

Results.

Parameters	Estimated values
θ_1	358.0
θ_2	-1326.
θ_3	301.5
σ^2	6.802

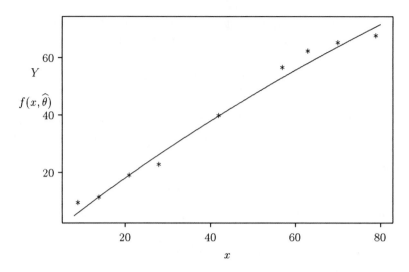

FIGURE 4.1. Pasture regrowth example: graph of observed and adjusted values of the response after estimation using a concave-shaped curve

Figure 4.1 shows that this choice of the regression function is wrong, because it badly estimates the decreasing speed; moreover, the estimated value of θ_1, $\hat{\theta}_1 = 358$ is not in accordance with the expected value (the graphic of observed values suggests the presence of an upper asymptote at a value of yield close to 70). Thus, we reject this regression function.

4.2.2 Isomerization example: graphics for diagnostic

Consider the graph of fitted versus observed values drawn in figure 1.9. To make this graph easier to read, one can plot the first diagonal and then superimpose a curve joining the points after smoothing (see figure 4.2).

Model. The regression function is

$$f(x,\theta) = \frac{\theta_1\theta_3(P - I/1.632)}{1 + \theta_2H + \theta_3P + \theta_4I},$$

where x is a three-dimensional variate, $x = (H, P, I)$, and the variances are homogeneous: $\text{Var}(\varepsilon_i) = \sigma^2$.

Method. The parameters are estimated by minimizing the sum of squares, $C(\theta)$; see equation (1.10).

The plot shown in figure 4.2 does not suggest any model misspecification: the points are well distributed around the first diagonal, and the smoothed scatter plot is close to the first diagonal.

Several plots of residuals also are shown in figure 4.3. None of these presents a particular structure; thus, we have no reason to suspect a departure from the hypotheses.

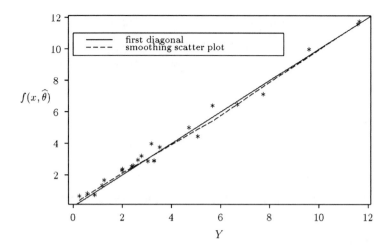

FIGURE 4.2. Isomerization example: graph of adjusted versus observed values

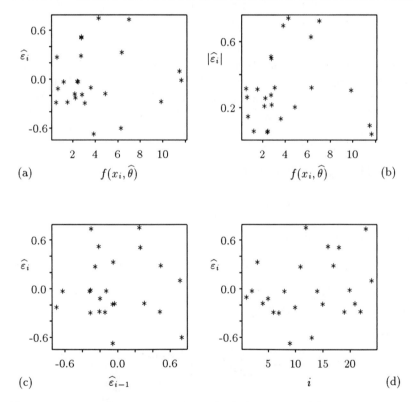

FIGURE 4.3. Isomerization example: residual plots. Figure (a) plots the residuals versus the fitted reaction rate. Figure (b) plots the absolute residuals versus the fitted reaction rate. Figure (c) plots the residuals versus the preceding residual. Figure (d) plots the residuals versus the residual order.

4.2.3 Peptides example: graphics for diagnostic

Let us consider the example described in section 3.1.2, assuming a logistic curve for the regression function (see equation (3.2)) and a constant variance.

Model. The regression function is

$$f(x, \theta) = \frac{100}{1 + \exp(\theta_2 (x - \theta_1))},$$

and the variances are homogeneous: $\text{Var}(\varepsilon_i) = \sigma^2$.

Method. The parameters are estimated by minimizing the sum of squares, $C(\theta)$; see equation (1.10).

Results.

Parameters	Estimated values	Estimated standard error
θ_1	43.92	0.792
θ_2	0.2052	0.0316
σ^2	328.4	

Figure 4.4 shows that the response is overestimated for fitted values lower than 50 (see figures 4.4 (a) and (b)). The plot of absolute residuals versus fitted values of the regression function shows that the variance of errors is smaller for values of the response close to 0 or 100.

Let us consider now the model defined in section 3.1.2, taking into account the heteroscedasticity using equation (3.3).

Model. The regression function is

$$f(x, \theta) = \frac{100}{1 + \exp(\theta_2(x - \theta_1))}, \tag{4.2}$$

and the variances are heterogeneous:

$$g(x, \sigma^2, \theta, \tau) = \sigma^2 + \sigma^2 \tau f(x, \theta) \left[100 - f(x, \theta)\right]. \tag{4.3}$$

Method. The parameters are estimated by maximizing the log-likelihood, $V(\theta, \sigma^2, \tau)$; see equation (3.8).

Results.

Parameters	Estimated values	Estimated standard error
θ_1	43.93	0.628
θ_2	0.4331	0.0557
τ_1	0.02445	0.00897
σ^2	26.79	

Figure 4.5 shows that the results of the estimation differ from the preceding case when we assume homogeneous variances. The estimated values

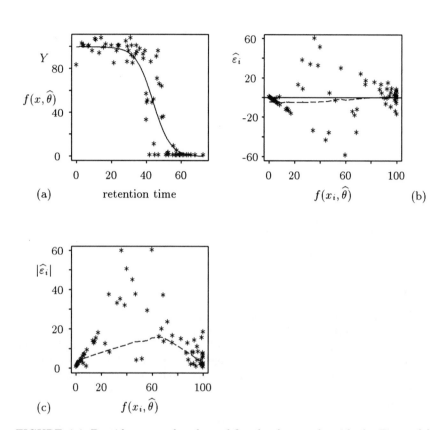

FIGURE 4.4. Peptides example: plots of fitted values and residuals. Figure (a) represents the observed (using a ∗) and fitted (using a line) values of solubility versus retention time. Figure (b) shows the residuals versus the fitted; the solid line is horizontal, passing through 0, and the dotted line is a curve joining the points after smoothing. Figure (c) represents the absolute residuals versus fitted values; the dotted line is a curve joining the points after smoothing.

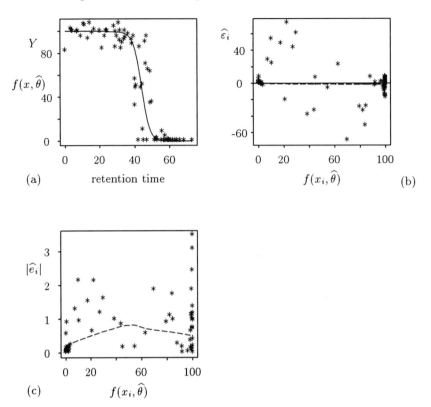

FIGURE 4.5. Peptides example: plots of fitted values and residuals taking into account heteroscedasticity. Figure (a) represents the observed and fitted values of solubility versus retention time. Figure (b) shows the residuals versus the fitted; the solid line is horizontal, passing through 0 and the dotted line is a curve joining the points after smoothing. Figure (c) represents the absolute standardized residuals versus fitted values; the dotted line is a curve joining the points after smoothing.

of the response are in better accordance with the observed values; see figure 4.5 (a) and (b). The estimated value of θ_2 is now equal to 0.4331, in place of 0.2052. Thus, the estimation of the maximum of the slope has increased, and the problem of overestimation for values of the response lower than 50 has disappeared. The interpretation of figure 4.5 (c) showing the absolute values of the standardized residuals versus the fitted responses is not so clear, however; if the variance function were well modeled, the points of this graph would not present any particular structure. We can only say that the structure of the points is not as pronounced as before, when we assumed a constant variance. We will return to this example in section 4.4.

4.2.4   Cortisol assay example: how to choose the variance function using replications

Figure 1.3 suggests that we can model the relationship between the response and the log-dose by using a symmetric sigmoidally shaped curve:

$$f(x,\theta) = \theta_1 + \frac{\theta_2 - \theta_1}{1 + \exp(\theta_3 + \theta_4 x)}.$$

This equation is similar to equation (1.3) with $\theta_5 = 1$. To illustrate a frequent mistake, let us see what happens if we hypothesize that the variance is constant rather than heteroscedastic.

Model. The regression function is

$$f(x,\theta) = \theta_1 + \frac{\theta_2 - \theta_1}{1 + \exp(\theta_3 + \theta_4 x)},$$

and the variances are homogeneous: $\mathrm{Var}(\varepsilon_i) = \sigma^2$.

Method. The parameters are estimated by minimizing the sum of squares, $C(\theta)$; see equation (1.10).

Results.

Parameters	Estimated values	Estimated standard error
θ_1	175.46	19.1
θ_2	2777.3	18.1
θ_3	2.0125	.065
θ_4	2.6872	.068
σ^2	3085	

The graph of absolute residuals versus fitted values of the response (see figure 4.6) clearly shows that the dispersion of the residuals varies with the values of the fitted response. This suggests that $\mathrm{Var}\varepsilon_{ij} = \sigma_i^2$.

In this example, the experimental design allows us to calculate the s_i^2 see equation (1.4), which estimate $\mathrm{Var}Y_{ij}$ independently of the estimated regression function. Graphs of s_i^2, or functions of s_i^2 versus $Y_{i\bullet}$, can be used to choose the variance function g or to confirm the presence of heteroscedasticity, as we see in figure 4.7. The graphs presented in figure 4.7 show that the relation between s_i and $Y_{i\bullet}$ is nearly linear (see figure 4.7 (b)), thereby suggesting that the variance of the response could be assumed proportional to the squared response. It follows that the function $g(x_i, \sigma^2, \theta) = \sigma^2 f^2(x_i, \theta)$ is a good candidate for the variance function. Another choice could be $g(x_i, \sigma^2, \theta, \tau) = \sigma^2 f^\tau(x_i, \theta)$, where τ is a parameter to be estimated; see chapter 3.

Now, let us again estimate the parameters, this time under the assumption $\mathrm{Var}Y_{ij} = \sigma^2 f^2(x_i, \theta)$.

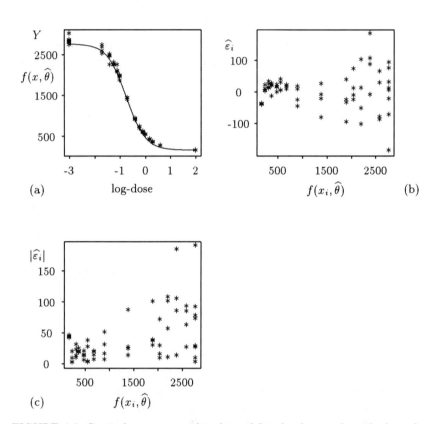

FIGURE 4.6. Cortisol assay example: plots of fitted values and residuals under the hypothesis of a symmetric sigmoidally shaped regression function and of a constant variance. Figure (a) represents the observed and fitted values of counts versus the log-dose. Figure (b) shows the residuals versus the fitted. Figure (c) represents the absolute residuals versus fitted values.

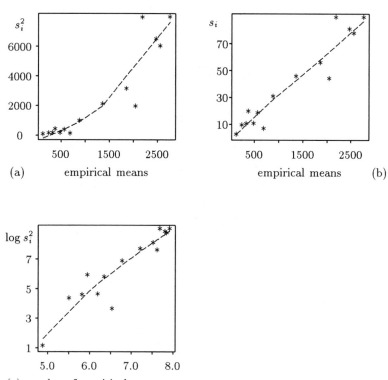

FIGURE 4.7. Cortisol assay example: plots of empirical variances versus empirical means. A curve joining the points after smoothing (broken line) is superimposed on each plot. Figure (a) represents the empirical variances versus the empirical means. Figure (b) represents the square-root of the empirical variances versus the empirical means. Figure (c) represents the logarithm of the empirical variances versus the logarithm of empirical means.

Model. The regression function is

$$f(x_i, \theta) = \theta_1 + \frac{\theta_2 - \theta_1}{1 + \exp(\theta_3 + \theta_4 x_i)}, \tag{4.4}$$

and the variances are heterogeneous:

$$\mathrm{Var}(\varepsilon_{ij}) = \sigma^2 f^2(x_i, \theta).$$

Method. The parameters are estimated by maximizing the log-likelihood, $V(\theta, \sigma^2, \tau)$; see equation (3.8).

Results.

Parameters	Estimated values	Estimated standard error
θ_1	137.15	2.64
θ_2	2855.8	33.9
θ_3	1.8543	.021
θ_4	2.3840	.034
σ^2	0.001614	

From the plots presented in figure 4.8, we can see that the points on the graph of absolute standardized residuals versus fitted values of the response (figure 4.8 (c)) are nearly uniformly distributed. This confirms the choice of the variance function. On the other hand, the plot of standardized residuals shows that the response is overestimated for low values (between 500 and 1500) and underestimated between 1500 and 2500. This result suggests that the maximum of decrease is closer to the upper asymptote, and it points us to the generalized logistic model in equation (1.3).

Model. The regression function is

$$f(x_i, \theta) = \theta_1 + \frac{\theta_2 - \theta_1}{(1 + \exp(\theta_3 + \theta_4 x_i))^{\theta_5}}, \tag{4.5}$$

and the variances are heterogeneous:

$$\mathrm{Var}(\varepsilon_{ij}) = \sigma^2 f^2(x_i, \theta).$$

Method. The parameters are estimated by maximizing the log-likelihood, $V(\theta, \sigma^2, \tau)$; see equation (3.8).

Results.

Parameters	Estimated values	Estimated standard error
θ_1	133.42	1.94
θ_2	2758.7	26.3
θ_3	3.2011	.223
θ_4	3.2619	.159
θ_5	0.6084	.041
σ^2	0.0008689	

The graphics presented in figure 4.9 do not suggest any model misspecification.

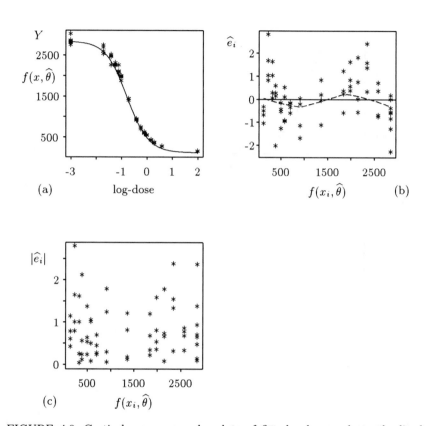

FIGURE 4.8. Cortisol assay example: plots of fitted values and standardized residuals under the hypothesis of a symmetric sigmoidally shaped regression function and heteroscedasticity. Figure (a) represents the observed and fitted values of counts versus the log-dose. Figure (b) shows the standardized residuals versus the fitted. A curve joining the points after smoothing (broken line) is superimposed. Figure (c) represents the absolute standardized residuals versus fitted values.

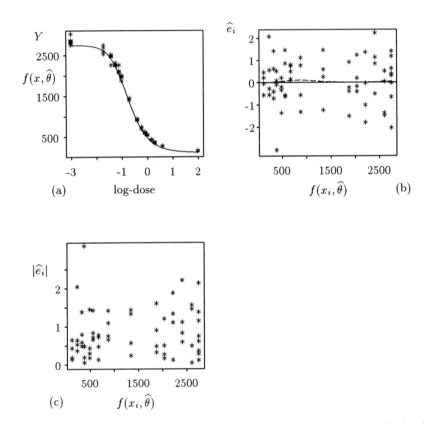

FIGURE 4.9. Cortisol assay example: plots of fitted values and standardized residuals under the hypothesis of an asymmetric sigmoidally shaped regression function and heteroscedasticity. Figure (a) represents the observed and fitted values of counts versus the log-dose. Figure (b) shows the standardized residuals versus the fitted. A curve joining the points after smoothing (broken line) is superimposed. Figure (c) represents the absolute standardized residuals versus fitted values.

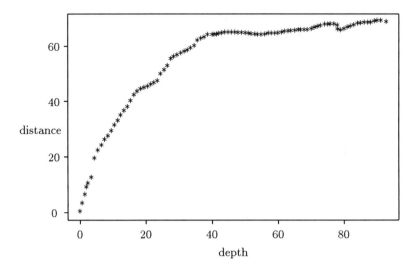

FIGURE 4.10. Trajectory of roots of maize: distance versus depth

4.2.5 Trajectory of roots of maize: how to detect correlations in errors

When correlations exist between the observations, the graphic of the residuals $\widehat{\varepsilon}_i$ versus $\widehat{\varepsilon}_{i-1}$ is a good tool to detect this lack of independence. Let us introduce a new example to illustrate this case.

The trajectory of roots is projected onto a vertical plane; the scientist then draws this projection to create a data set by numerization [TP90]. The variable Y is the distance between the root and the vertical axis running through the foot of the plant, and x is the depth in the ground. The scientist can pick up as many points as necessary to describe the trajectory of each root, as shown in figure 4.10.

A nonlinear regression model describes this phenomenon by modeling the relation between Y and x with the function $f(x, \theta) = \theta_1(1 - \exp(-\theta_2 x))$.

Model. The regression function is

$$f(x, \theta) = \theta_1(1 - exp(-\theta_2 x)),$$

and the variances are homogeneous: $\text{Var}(\varepsilon_i) = \sigma^2$.

Method. The parameters are estimated by minimizing the sum of squares, $C(\theta)$; see equation (1.10).

Results.

Parameters	Estimated values
θ_1	67.876
θ_2	0.05848

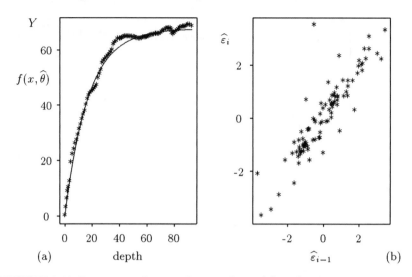

(a)   (b)

FIGURE 4.11. Trajectory of roots of maize: figure (a) is the plot of observed and adjusted values of the response; figure (b) is the plot of residual versus preceding residual.

Figure 4.10 clearly shows that the quantities $Y_i - f(x_i, \theta)$ and $Y_j - f(x_j, \theta)$ are not independent when i and j are close to each other. Moreover, the points of the graph (see figure 4.11 (b)) of $\widehat{\varepsilon}_i$ versus $\widehat{\varepsilon}_{i-1}$ are distributed along the first diagonal. This distribution suggests that the correlation between the observations could be modeled by an autoregressive model: $\varepsilon_i = \rho \varepsilon_{i-1} + \eta_i$, where $\rho \in [-1, 1]$ is a parameter, generally unknown, and η_i for $i = 1, \ldots n$ are independent random variables, with expectation 0 and variance σ^2. Models taking into account correlated observations are not treated in this book, but interested users will find several publications on that subject, for example: Glasbey [Gla80] and White and Domowitz [WD84].

4.2.6   What can we say about the experimental design?

Obvious links exist between the choice of the regression function and the choice of the experimental design, $x_1, x_2, \ldots x_k$. For example, k must be greater than p.

Consider again the cortisol example. Intuitively, we know that if we suppress the observations corresponding to a logarithm of the dose between -1 and 0, the parameters θ_3, θ_4 and θ_5 will not be estimated *easily*: either the estimating procedure will not converge, or the estimated standard error of the estimated parameters will be very large with respect to their values, or the estimated values of the parameters will be aberrant for the scientist.

In this case, one must verify if the values of x and the regression function are in accord. For this purpose we use sensitivity functions to describe the link between f and x.

Let $f(x, \theta)$ be the regression function, where θ is the vector with p components θ_a, $a = 1, \ldots p$. The sensitivity function for parameter θ_a is the derivative of f with respect to θ_a:

$$\Phi_a(x, \theta) = \frac{\partial f}{\partial \theta_a}(x, \theta).$$

Consider that the value of x is fixed; thus, if the absolute value of $\Phi_a(x, \theta)$ is small, the variations of $f(x, \theta)$ are small when θ_a varies. It follows that in order to estimate θ_a accurately, we must choose x such that the absolute value of $\Phi_a(x, \theta)$ is large.

Let us illustrate this procedure with a simulated data set.

Data. The data are simulated in the following way. Let $\mu(x) = 1 - \exp(-x)$, and let ε be a centered normal variable with variance $25 \, 10^{-4}$. The values of x are chosen in the following way: $x_i = i/10$ for $i = 1, \ldots 10$. The observations Y_i are defined as $Y_i = \mu(x_i) + \varepsilon_i$, where the ε_i, $i = 1, \ldots 10$, are independent variables with the same distribution as ε.

x	0.1	0.2	0.3	0.4	0.5
Y	0.1913	0.0737	0.2702	0.4270	0.2968
x	0.6	0.7	0.8	0.9	1.0
Y	0.4474	0.4941	0.5682	0.5630	0.6636

Model. The regression function is

$$f(x, \theta) = \theta_1(1 - \exp(-\theta_2 x)),$$

and the variances are homogeneous: $\mathrm{Var}(\varepsilon_i) = \sigma^2$.

Method. The parameters are estimated by minimizing the sum of squares, $C(\theta)$; see equation (1.10).

Results.

Parameters	Estimated values	Estimated standard errors
θ_1	1.1007	0.49
θ_2	0.8654	0.54
σ^2	0.00418	

The estimated standard errors are very large. Using formula (2.3), we find that the confidence interval for θ_1, with asymptotic level 95%, is $[0.13; 2.07]$, and for θ_2 it is $[-0.19; 1.93]$. The interval for θ_2 contains 0, which is an unacceptable value in this example.

Looking at figure 4.12, we can posit that the estimates would be more accurate with a wider interval of variations for x. To quantify this remark, let us look at the plot of sensitivity functions.

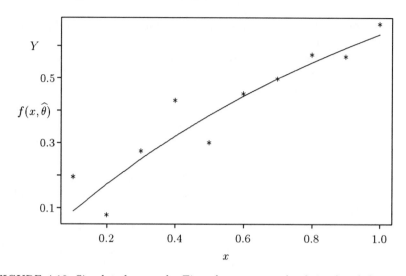

FIGURE 4.12. Simulated example: First data set, graph of simulated data and fitted values of the response

The functions Φ_a are defined as follows:

$$
\begin{aligned}
\Phi_1(x,\theta) &= 1 - \exp(-\theta_2 x) \\
\Phi_2(x,\theta) &= \theta_1 x \exp(-\theta_2 x).
\end{aligned}
$$

Generally, the value of θ is unknown and the functions Φ_a are calculated in $\widehat{\theta}$.

Figure 4.13 shows the plot of sensitivity functions $\Phi_1(x,\widehat{\theta})$ and $\Phi_2(x,\widehat{\theta})$. We see that median values of x, let us say between 0.5 and 2, contribute to the estimation of θ_2 with high accuracy, while greater values of x contribute to the estimation of θ_1.

Now, let us see what happens with more appropriate values of x.

Data. The data now are simulated in the following way. The values of x are chosen in the interval $[.1; 4.6]$. The observations Y_i are defined as $Y_i = \mu(x_i) + \varepsilon_i$, where the ε_i, $i = 1, \ldots 10$, are independent variables with the same distribution as ε.

x	0.1	0.6	1.1	1.6	2.1
Y	0.1913	0.3436	0.6782	0.8954	0.7809
x	2.6	3.1	3.6	4.1	4.6
Y	0.9220	0.9457	0.9902	0.9530	1.0215

Model. The regression function is

$$
f(x,\theta) = \theta_1(1 - \exp(-\theta_2 x)),
$$

and the variances are homogeneous: $\mathrm{Var}(\varepsilon_i) = \sigma^2$.

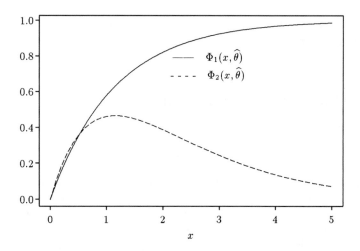

FIGURE 4.13. Simulated example: graph of sensitivity functions calculated in $\widehat{\theta}$ versus x

Method. The parameters are estimated by minimizing the sum of squares, $C(\theta)$; see equation (1.10).

Results.

Parameters	Estimated values	Estimated standard errors
θ_1	1.1006	0.04
θ_2	0.9512	0.14

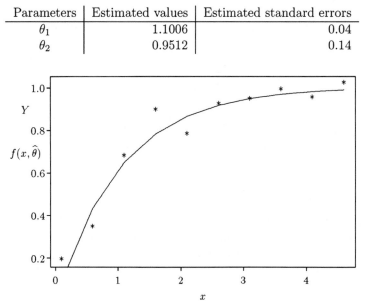

FIGURE 4.14. Simulated example: second data set, graph of simulated data and fitted value of the response

As expected, the estimated standard errors are smaller than they were in the first data set. Comparing figures 4.14 and 4.12 clearly shows that the experimental design in the second data set allows us to estimate the asymptote θ_1 better.

4.3   Diagnostics of model misspecifications with tests

4.3.1   RIA of cortisol: comparison of nested models

In some situations it is easy to find possible alternatives to the regression function or the variance function. For example, if we choose a symmetric sigmoidally shaped curve for the cortisol data, it is reasonable to wonder if an asymmetric curve would not be more appropriate.

Let us consider the models defined in equations (4.4) and (4.5). If we suspect an asymmetry of the curve, we can test the hypothesis H: "$\theta_5 = 1$" against the alternative A: "$\theta_5 \neq 1$." Let us apply a likelihood ratio test. Let $\widehat{\theta}_H$ be the estimator of θ under the hypothesis H and $\widehat{\sigma}_H^2 f^2(x_i, \widehat{\theta}_H)$ be the estimator of $\mathrm{Var}(\varepsilon_{ij})$. In the same way, let us define $\widehat{\sigma}_A^2 f^2(x_i, \widehat{\theta}_A)$ as the estimator of $\mathrm{Var}(\varepsilon_{ij})$ under the alternative A. The test statistic

$$S_L = \sum_{i=1}^{k} \log \widehat{\sigma}_H^2 f^2(x_i, \widehat{\theta}_H) - \sum_{i=1}^{k} \log \widehat{\sigma}_A^2 f^2(x_i, \widehat{\theta}_A)$$

equals 39, which must be compared to the 0.95 quantile of a χ^2 with 1 degree of freedom. The hypothesis of a symmetric regression function is rejected.

The following section presents a likelihood ratio type test based on the comparison between the considered model and a "bigger" model defined when the response is observed with replications.

4.3.2   Tests using replications

Let us return to the general nonlinear regression model with replications. We assume that for each value of x, x_i, $i = 1, \ldots k$, the number n_i of observed values of the response is large, greater than 4 for example. We choose a regression function, f, and carry out the estimation of the parameters. Let H be the considered model:

$$\left. \begin{array}{rcl} Y_{ij} &=& f(x_i, \theta) + \varepsilon_{ij} \\ \mathrm{Var}(\varepsilon_{ij}) &=& \sigma^2, \; \mathrm{E}(\varepsilon_{ij}) = 0 \end{array} \right\}.$$

$\widehat{\theta}_H$ is the estimation of θ and $\widehat{\sigma}_H^2$ the estimation of σ^2. Under H, we estimate $p + 1$ parameters.

H can be considered as a model nested in a more general one, denoted by A_1, defined in the following way:

$$\left.\begin{array}{rcl} Y_{ij} & = & \mu_i + \varepsilon_{ij} \\ \mathrm{Var}(\varepsilon_{ij}) & = & \sigma^2, \ \mathrm{E}(\varepsilon_{ij}) = 0 \end{array}\right\}.$$

Now the parameters to be estimated are $\mu_1, \dots \mu_k$ and σ^2, that is, $k+1$ parameters. The presence of replications in each x_i allows us to estimate μ_i by the empirical mean $Y_{i\bullet}$ and to build a test of the hypothesis H against the alternative A_1 as in section 2.3.3 or 3.4.2. The test statistic

$$\mathcal{S}_\mathrm{L} = n \log C(\widehat{\theta}_\mathrm{H}) - n \log \sum_{i=1}^k n_i s_i^2$$

is asymptotically distributed as a χ^2 with $k - p$ degrees of freedom.

If the test is not significant, we have no reason to reject the regression function f.

By modifying A_1, one may also test the hypothesis of homogeneous variances. Let A_2 be defined by:

$$\left.\begin{array}{rcl} Y_{ij} & = & \mu_i + \varepsilon_{ij} \\ \mathrm{Var}(\varepsilon_{ij}) & = & \sigma_i^2, \ \mathrm{E}(\varepsilon_{ij}) = 0 \end{array}\right\}.$$

Under the model A_2, we estimate $2k$ parameters. The μ_i are estimated by the empirical means, and the σ_i^2 are estimated by the empirical variances, denoted by s_i^2. The test statistic

$$\mathcal{S}_\mathrm{L} = n \log \widehat{\sigma}_\mathrm{H}^2 - \sum_{i=1}^k n_i \log s_i^2,$$

where $\widehat{\sigma}_\mathrm{H}^2 = C(\widehat{\theta}_\mathrm{H})/n$, is asymptotically distributed as a χ^2 with $2k - p - 1$ degrees of freedom.

This test is easily extended to the case of heterogeneous variances under the model H. Let the considered model H be the following:

$$\left.\begin{array}{rcl} Y_{ij} & = & f(x_i, \theta) + \varepsilon_{ij} \\ \mathrm{Var}(\varepsilon_{ij}) & = & g(x_i, \sigma^2, \theta, \tau), \ \mathrm{E}(\varepsilon_{ij}) = 0 \end{array}\right\}.$$

Under H, we estimate $p + q + 1$ parameters (p for θ, q for τ and 1 for σ^2). The test statistic

$$\mathcal{S}_\mathrm{L} = \sum_{i=1}^k n_i \log \widehat{\sigma}_\mathrm{H}^2 g(x_i, \widehat{\theta}_\mathrm{H}) - \sum_{i=1}^k n_i \log s_i^2$$

is asymptotically distributed as a χ^2 with $2k - p - q - 1$ degrees of freedom.

We illustrate this method based on replications using the cortisol assay example and the ovocytes example.

4.3.3 Cortisol assay example: misspecification tests using replications

Let us consider again the model H where the regression function is assumed symmetric and the variance is assumed proportional to the squared expectation.

Model. The regression function is defined by equation (4.4), and the variances are heterogeneous:

$$\text{Var}(\varepsilon_{ij}) = \sigma^2 f^2(x_i, \theta).$$

Method. The parameters are estimated by maximizing the log-likelihood, $V(\theta, \sigma^2, \tau)$; see equation (3.8).

Results. We test the model H against the alternative A_2. The test statistic S_L equals 50. This number must be compared to 37, the 0.95 quantile of a χ^2 with 25 ($k = 15$, and the number of parameters to be estimated is 5, 4 for θ and 1 for σ^2) degrees of freedom. The hypothesis H is rejected.

Let us see what happens if we take into account an asymmetry in the regression function.

Model. The regression function is defined by equation (4.5), and the variances are heterogeneous:

$$\text{Var}(\varepsilon_{ij}) = \sigma^2 f^2(x_i, \theta).$$

Method. The parameters are estimated by maximizing the log-likelihood, $V(\theta, \sigma^2, \tau)$; see equation (3.8).

Results. The test statistic S_L equals 10.5, which must be compared to 36.4, the 0.95 quantile of a χ^2 with 24 degrees of freedom. The hypothesis H is not rejected. The analysis of graphics of residuals shown in figure 4.8 is strengthened.

4.3.4 Ovocytes example: graphics of residuals and misspecification tests using replications

Let us return to the ovocytes example already treated in section 1.4.4. Let H be the following model.

Model. The regression function is

$$f(t, P_w, P_s) = \frac{1}{V_0}(V_w(t) + V_s(t) + V_x),$$

where V_w and V_s are solutions of equation (1.8), and the variances are homogeneous: $\text{Var}(\varepsilon_i) = \sigma^2$.

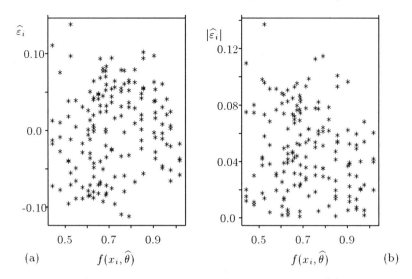

FIGURE 4.15. Ovocytes example: plots of residuals. Figure (a) is the plot of residuals versus fitted volume. Figure (b) is the plot of absolute standardized residuals versus fitted volume

Method. The parameters are estimated by minimizing the sum of squares, $C(\theta)$; see equation (1.10).

Results. Figure 4.15 shows the plots of residuals. The plots do not show any particular structure, except for a weak decrease of the absolute residuals when plotted against fitted values of the regression function. This behavior seems to be confirmed by the graph illustrating empirical variances versus empirical means (see figure 4.16). However, this graph also shows that the empirical variances vary a lot.

Let us now see the result of a misspecification test. The test statistic of hypothesis H against A_2, S_L, equals 31, which must be compared to 71, the value of the 0.95 quantile of a χ^2 with 53 degrees of freedom (in that example, $k = 29$). Thus there is no reason to reject the hypothesis of a constant variance.

4.4 Numerical troubles during the estimation process: peptides example

In this section we tackle a tedious problem: what to do when faced with numerical problems during the estimation process? In nonlinear regression models, estimating the parameters needs an iterative numerical process. If the parameters are estimated by the least squares method, the numerical process minimizes the sum of squares $C(\theta)$ in equation (1.10); if the parameters are estimated by maximizing the log-likelihood $V(\theta, \sigma^2, \tau)$ in

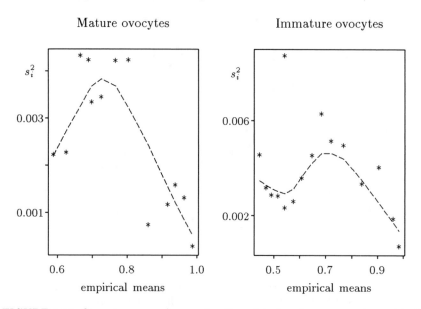

FIGURE 4.16. Ovocytes example: graphs of empirical variances versus empirical means. A curve joining the points after smoothing (broken line) is superimposed on each plot

equation (3.8), the numerical process minimizes $-2V(\theta, \sigma^2, \tau)$. Starting from some initial values for the parameters to be estimated, the program calculates the criterion to minimize, and, using a specific algorithm (see for example [SW89]), it looks for new values of the parameter for which the value of the criterion to minimize is smaller. This procedure is stopped when a stopping criterion is small enough.

Generally, the user does not intervene in this iterative estimation process, because the methods used and their numerical options are automatically chosen by the program. Nevertheless, he has to choose the initial values of the parameters to start the iterative process.

All of the examples discussed in our book are treated using **nls2**, and we did not have numerical difficulties for any of them during the iterative process for estimating the parameters. But as all users of nonlinear regression models know, numerical problems may appear, and, generally, the software's output is not informative enough to help and leaves the user puzzled.

Although it is difficult to list all of reasons leading to a numerical problem, let us try to give some troubleshooting procedures.

1. Check the data set and the model.

2. Try to run the program again with another set of initial values.

3. Change the method used to estimate the parameters. This suggestion

applies only in the case of heterogeneous variances when the parameters are estimated by minimizing $-2 \log V(\theta, \sigma^2, \tau)$. If the iterative process fails to converge, then the user may run the program again, choosing another estimator, like a three-step alternate least squares (see section (3.3.2)).

4. Change the parameterization of the regression function or the variance function. Sometimes, looking carefully at the behavior of the iterative estimation process, it appears that the problem comes from the estimation of one or two parameters. In some cases, the estimation process may run better by changing the parameterization of the model. For example, the following functions differ only by their parameterization:

$$
\begin{aligned}
f(x, \theta) &= \exp\left(\theta_2(x - \theta_1)\right) \\
f(x, \alpha) &= \exp(\alpha_1 + \alpha_2 x), \text{ with } \alpha_1 = -\theta_1\theta_2 \text{ and } \alpha_2 = \theta_2 \\
f(x, \beta) &= \beta_1 \exp \beta_2 x, \text{ with } \beta_1 = \exp(-\theta_1\theta_2) \text{ and } \beta_2 = \theta_2 \\
f(x, \gamma) &= \gamma_1 \gamma_2^x, \text{ with } \gamma_1 = \exp(-\theta_1\theta_2) \text{ and } \gamma_2 = \exp \theta_2.
\end{aligned}
$$

5. Change the model. For example, we chose the Weibull model to represent a sigmoidally shaped curve in the pasture regrowth example, but several other models are available:

$$
\begin{aligned}
f_1(x, \theta) &= \theta_1 \exp\left(\theta_2/(x + \theta_3)\right) \\
f_2(x, \theta) &= \theta_1 \exp\left(-\exp(\theta_2 - \theta_3 x)\right) \\
f_3(x, \theta) &= \theta_1 + \theta_2/\left(1 + \exp(\theta_3 - \theta_4 x)\right).
\end{aligned}
$$

A list of nonlinear regression models can be found in the book by Ratkowsky [Rat89].

Let us illustrate some of these remarks with our peptides example by examining again the choice of the variance function. To take into account the fact that the variability is greater when the peptide is fully soluble than when it is nonsoluble, we propose to generalize equation (4.3) to the following:

$$
\sigma_i^2 = \sigma^2 + \sigma^2 \tau_1 f(x_i, \theta)\left[100 + \tau_2 - f(x_i, \theta)\right] = g\left(x_i, \theta, \sigma^2, \tau_1, \tau_2\right).
$$

When x equals 0, $g\left(x_i, \theta, \sigma^2, \tau_1, \tau_2\right)$ equals $\sigma^2(1 + 100\tau_1\tau_2)$, and when x tends to infinity, $g\left(x_i, \theta, \sigma^2, \tau_1, \tau_2\right)$ tends to σ^2. Let us estimate the parameters with this new model.

Model. The regression function is

$$
f(x, \theta) = \frac{100}{1 + \exp\left(\theta_2 (x - \theta_1)\right)},
$$

and the variance function is

$$
\text{Var}(\varepsilon_i) = \sigma^2(1 + \tau_1 f(x_i, \theta)(100 + \tau_2 - f(x_i, \theta))). \tag{4.6}
$$

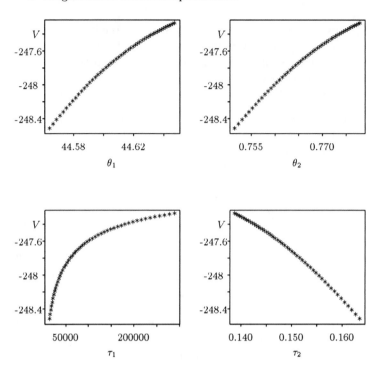

FIGURE 4.17. Peptides example: variations of the likelihood against the parameters' values during the estimation process. For each parameter to be estimated, the values of the parameter at iterations $250, 255, \ldots 500$ are reported on the horizontal axis, the corresponding values of the log-likelihood being reported on the vertical axis

Method. The parameters are estimated by maximizing the log-likelihood, $V(\theta, \sigma^2, \tau)$; see equation (3.8).

Results. We start the estimation process using for initial parameter values the estimations of θ and τ_1 calculated with the model defined by equations (4.2) and (4.3) and $\tau_2 = 0$. We find the following results after 500 iterations of the iterative process:

	Initial values	Estimated values	Estimated standard error
θ_1	43.93	44.65	0.430
θ_2	0.433	0.778	0.0393
τ_1	0.024	300898	106278
τ_2	0	0.138	0.0061

The estimation process did not converge because the variations of the likelihood are very slow even for very big variations of τ_1. Figure 4.17 shows the variations of V versus the parameter values obtained at iterations $250, 255, \ldots 500$. It appears that when τ_1 varies from 17000 to 270000, V varies from -247.8 to -247.3.

Let us try to run the estimation process with another model for the variance function defined as follows:

$$\sigma_i^2 = \sigma^2 \exp\left(\tau_1 f(x_i, \theta)(100 + \tau_2 - f(x_i, \theta))\right) = g\left(x_i, \theta, \sigma^2, \tau_1, \tau_2\right).$$

This function is obviously different from the preceding polynomial variance function, but the curve is concave and looks like a bell-shaped curve. Moreover, it is often preferable to model a variance function as a function of the exponential function because the result is always positive. Let us see the results with this new model.

Model. The regression function is

$$f(x, \theta) = \frac{100}{1 + \exp\left(\theta_2\left(x - \theta_1\right)\right)},$$

and the variance function is

$$\text{Var}(\varepsilon_i) = \sigma^2 \exp\left(\tau_1 f(x_i, \theta)(100 + \tau_2 - f(x_i, \theta))\right). \tag{4.7}$$

Method. The parameters are estimated by maximizing the log-likelihood, $V(\theta, \sigma^2, \tau)$; see equation (3.8).

Results. We start the estimation process using for initial parameter values the estimations of θ calculated with the model defined by equations (4.2) and (4.3) and $\tau_2 = 0$. To choose the initial value for τ_1, we compare the maximum in f of the function $\exp(\tau_1 f(100 - f))$, equal to $\exp 50^2 \tau_1$, to the maximum of the function $1 + \tau_1 f(100 - f)$, equal to $1 + 50^2 \tau_1$. Thus $\tau_1 = 50^{-2} \log(1 + 50^2 \times 0.024)$ is a possible initial value for τ_1. We take $\tau_1 = 0.001$.

	Initial values	Estimated values	Estimated standard error
θ_1	43.93	45.07	0.422
θ_2	0.433	0.327	0.0223
τ_1	0.001	0.0025	0.00018
τ_2	0	10.45	1.41
σ^2		2.766	

After 49 iterations, the estimation process did converge. Thus, in this particular case we were able to circumvent the numerical difficulty by using another variance function. The shape of this function is nearly similar to the shape of the initial variance function defined by equation (4.6). Let us conclude this chapter by pursuing the discussion about the peptides example.

4.5   Peptides example: concluded

Figure 4.18 presents the graph of fitted values, residuals and standardized absolute residuals when we use the exponential function to model the variations of the variance. Figures 4.18 (a) and (b) show that the response is

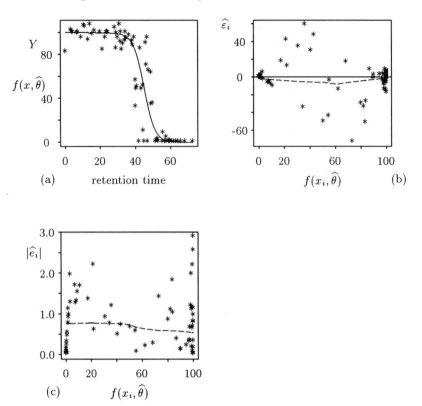

FIGURE 4.18. Peptides example: graphs of fitted values and residuals using the exponential variance function. Figure (a) represents the observed and fitted values of the solubility versus the retention time. Figure (b) shows the residuals versus the fitted; the solid line is the horizontal line passing through 0, and the dotted line is a curve joining the points after smoothing. Figure (c) represents the absolute standardized residuals versus fitted values; the dotted line is a curve joining the points after smoothing.

overestimated on its set of variation. Looking at figure 4.18 (c), it appears that the absolute standardized residuals versus fitted values do not present any particular structure.

Obviously, we cannot be satisfied with a bad estimation of the response curve. This poor fit is due to the choice of the variance function. Using model (4.7) has two effects. The first is, as expected, to take into account the bigger variability of the data when the peptide is fully soluble than when it is nonsoluble. With model (4.3), the fitted variance when $f = 0$ or $f = 100$ equals $\widehat{\sigma}^2 = 26.79$ (see the results given in section 4.2.3). With model (4.7), the fitted variance when $f = 0$ is $\widehat{\sigma}^2 = 2.77$, while it is $\widehat{\sigma}^2 \exp(100\widehat{\tau}_1\widehat{\tau}_2) = 38.9$ when $f = 100$. The second effect is to increase considerably the fitted variances when the response lies between 30 and 70, as is shown in figure 4.19. Thus, the weights allowed for observations

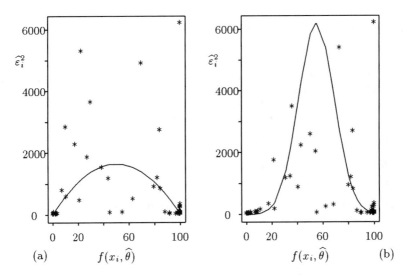

FIGURE 4.19. Pepdtides example: squared residuals $\widehat{\varepsilon}_i^2$ versus adjusted regression function $f(x_i, \widehat{\theta})$. The superimposed lines represent the fitted variances $g(x_i, \widehat{\theta}, \widehat{\sigma}^2, \widehat{\tau})$, versus the $f(x_i, \widehat{\theta})$. Figure (a) corresponds to variance function (4.3), Figure (b) to (4.7).

obtained for values of retention times between 40 and 50 are very low relative to the weights given for observations obtained for values of retention times close to 0 and 60. This explains the overestimation of the response.

4.6   Using **nls2**

This section reproduces the commands and files used in this chapter to analyze the examples using **nls2**.

Pasture regrowth example: parameter estimation and plotting the observed and fitted values of the response

Parameter estimation

In a file called `pasture.mod2`, we describe the model defined by equation (4.1), section 4.2.1, page 91:

```
resp yield;
varind time;
parresp p1, p2, p3;
subroutine;
begin
yield=(p2 + p1*time)/(time+p3);
end
```

The parameters are estimated by using the function nls2:

```
> pasture.nl2<-nls2(pasture,"pasture.mod2",c(70,-2,20))
>   # Print the results
> cat( "Estimated values of the parameters:\n ")
> print( pasture.nl2$theta)
> cat( "Estimated value of sigma2:\n ")
> print( pasture.nl2$sigma2); cat( "\n\n")
```

Plotting the observed and fitted values of the response

To plot the observed and fitted responses versus the independent variable, we use the function plfit (see figure 4.1, page 92):

```
> plfit(pasture.nl2, title ="Pasture regrowth example",
         sub = "Observed and adjusted response")
```

(Results for this example are given in paragraph 4.2.1, page 91.)

Isomerization example: graphics for diagnostics

The results of the estimation procedure by nls2 have been stored in the structure called isomer.nl1 (see section 1.6, page 27).

Plotting the fitted values of the response against its observed values

To plot the fitted values of the response against its observed values, we use the function plfit:
- option smooth means that the observed values are joined after smoothing.
- option wanted specifies the graph requested: fitted values versus observed values of the response.
- option ask.modify allows us to type an **S-PLUS** command before the graph disappears from the screen. Here, we type the abline command to add the first diagonal on the plot.

```
> plfit(isomer.nl1, smooth=T, wanted=list(O.F=T),
       title="Isomerization example",
       ask.modify=T)
> abline(0,1)
```

(See figure 4.2, page 93).

Plotting the residuals

The function plres is used to plot the residuals.
   Option wanted specifies the graphs requested. Its values are
- F.R: the residuals are plotted against the fitted values of the response,
- R.R: the residuals are plotted against the immediately preceding residuals,
- I.R: the residuals are plotted against their indexes.
   Option absolute means that the absolute values of the residuals are added on the graphs corresponding to F.R and I.R:

```
> plres(isomer.nl1,
      wanted=list(F.R=T, R.R=T, I.R=T),
      absolute=T,
      title="Isomerization example")
```

(See figure 4.3, page 93).

Solubility of peptides example: graphics for diagnostics

Plotting the observed and fitted values of the response when variance is assumed to be constant

The results of estimation with the model defined in section 3.1.2 (logistic curve for the regression function and constant variances) have been stored in the structure **pept.nl1** (see section 3.6, page 84). Now we plot the observed and fitted responses versus the retention time by using the function **plfit**. Option **wanted** specifies the graph requested. The observed and fitted values of the response are plotted against the independent variable:

```
> plfit(pept.nl1, wanted=list(X.OF=T),
        title="Solubility of peptides example - constant variance",
        sub="Observed and fitted response")
```

(See first plot, figure 4.4, page 95).

Plotting the residuals when variance is assumed to be constant

Residuals and absolute residuals are plotted against the fitted values of the response by using the function **plres**:
- option **wanted** specifies the graph requested: the fitted values of the residuals are plotted against the fitted values of the response.
- option **absolute** means that the absolute residuals are plotted.
- option **smooth** means that the observed values are joined after smoothing.
- option **ask.modify** allows us to type an **S-PLUS** command before the graph disappears from the screen. Here, we type the **abline** command to add an horizontal line on the plot.

```
> plres.nls2(pept.nl1,  wanted=list(F.R=T),
          absolute=T, smooth=T, ask.modify=T,
          title="Solubility of peptides example - constant variance")
>   abline(0,0)
```

(See the second and third plots, figure 4.4, page 95).

Plotting the observed and fitted values of the response and the residuals when variance is assumed to be heteroscedastic

In the same way, we draw plots from the structure **pept.nl3**, which contains the results of estimation when the variance is assumed to be heteroscedastic, (see section 3.6, page 84, and equation (3.3), page 65).

```
>    # Plot the observed and fitted values of the response
>    # against the independent variable.
> plfit(pept.nl3, wanted=list(X.OF=T),
        title="Solubility of peptides - non constant variance",
        sub="Observed and fitted response")
>    # Plot the residuals
> plres.nls2(pept.nl3, wanted=list(F.R=T),
        absolute=T, st=T, smooth=T, ask.modify=T,
        title="Solubility of peptides - non constant variance")
> abline(0,0)
```

(See figure 4.5, page 96).

Cortisol assay example: how to choose the variance function using replications

Estimation using a symmetric sigmoidally shaped regression curve and a constant variance

We first estimate the parameters by using the model defined in section 4.2.4, page 97 (symmetric regression curve and constant variance). When the parameter θ_5 is equal to 1, this model is similar to the model we described in the file corti.mod1 (asymmetric regression curve and constant variance, see section 1.6, page 21). So, by setting a numerical equality constraint on the fifth parameter, we can reuse the file corti.mod1 when calling nls2:

```
> corti.nl3<-nls2(corti,
    list(file="corti.mod1",gamf=c(0,10), eq.theta=c(rep(NaN,4),1)),
    c(corti.nl1$theta[1:4],1))
>    # Print the main results
> cat( "Estimated values of the parameters:\n ")
> print(coef(corti.nl3))
> cat( "Estimated value of sigma2:\n ")
> print( corti.nl3$sigma2); cat( "\n\n")
```

Plotting the response values and the residuals when the regression function is a symmetric sigmoidally shaped curve and the variance is constant

We use the graphical functions of **S-PLUS** to plot the observed and fitted values of the response versus the log-dose. The observed values are joined by a line. Then we plot the residuals and absolute residuals versus the fitted values of the response by using the function plres:

```
>    # Plot the observed and fitted values of the response
>    # against the log-dose
> plot(logdose,corti$cpm,xlab="log-dose",ylab="response")
> title(main="Cortisol example",
        sub="Observed and adjusted response")
> lines(unique(logdose),corti.nl3$response)
>    # Plot the residuals
> plres(corti.nl3,
```

```
wanted=list(F.R=T),
absolute=T,
title="Cortisol example")
```

(See figure 4.6, page 98).

Plotting the empirical variances versus the empirical means when the
regression function is a symmetric sigmoidally shaped curve and the
variance is constant

To plot the variances, we use the graphical function `plvar`. This function
offers many choices. By option `wanted`, we specify the graphs requested:
- `Y.S2` to plot the empirical variances against the empirical means,
- `Y.S` to plot the square roots of the empirical variances against the empirical means,
- `logY.logS2` to plot the logarithms of the empirical variances against the
logarithms of the means.

Option `smooth` means that the observed values are joined after smoothing:

```
> plvar(corti.nl3,wanted=list(Y.S2=T,Y.S=T,logY.logS2=T),
    smooth=T,
    title="Cortisol example")
```

(See figure 4.7, page 99).

Estimation using a symmetric sigmoidally shaped regression curve and a
heteroscedastic variance

Now we study the model defined on page 100 (symmetric sigmoidal regression curve and heteroscedastic variances).

We describe it in a file called `corti.mod6`. Actually, `corti.mod6` corresponds to an asymmetric model but, by setting the fifth parameter g to 1,
it is equivalent to a symmetric one. By doing so, we will not have to create
another file when we study the asymmetric model:

```
% model CORTI
resp cpm;
var v;
varind dose;
parresp n,d,a,b,g;
pbisresp minf,pinf;
subroutine;
begin
cpm= if dose <= minf then d else
     if dose >= pinf then n else
     n+(d-n)*exp(-g*log(1+exp(a+b*log10(dose))))
   fi fi;
v = cpm**2;
end
```

We estimate the parameters by using nls2, remembering to set a numerical constraint on the last parameter:

```
> corti.nl6<-nls2(corti,
   list(file= "corti.mod6",gamf=c(0,10), eq.theta=c(rep(NaN,4),1)),
   c(corti.nl1$theta[1:4],1))
>      # Print the main results
> cat( "Estimated values of the parameters:\n ")
> print(coef(corti.nl6))
> cat( "Estimated value of sigma2:\n ")
> print( corti.nl6$sigma2); cat( "\n\n")
```

Plotting the observed and fitted values of the response and the residuals when the regression function is a symmetric sigmoidally shaped curve and the variance is heteroscedastic

We plot the observed and fitted values of the response by using graphical functions of **S-PLUS** and the residuals by using the function plres:

```
>      # Plot the observed and fitted values of the response
>      # against the log-dose
> plot(logdose,corti$cpm,xlab="log-dose",ylab="response")
> title(main="Cortisol example",
         sub="Observed and adjusted response")
> lines(unique(logdose),corti.nl6$response)
>      # Plot the residuals
> plres(corti.nl6,
         wanted=list(F.R=T),
         absolute=T,st=T,smooth=T,
         title="Cortisol example",ask.modify=T)
> abline(0,0)
```

(See figure 4.8, page 101).

Estimation using a asymmetric sigmoidally shaped regression curve and an heteroscedastic variance

Finally, we study the last model defined on page 100 (asymmetric sigmoidal regression curve and heteroscedastic variances). It is similar to the previous one where we suppress the constraint on the last parameter. So, when calling nls2, we reuse the file corti.mod6:

```
> corti.nl7<-nls2(corti,
      list(file="corti.mod6",gamf=c(0,10)),
      corti.nl6$theta)
>      # Print the main results
> cat( "Estimated values of the parameters:\n ")
> print(coef(corti.nl7))
> cat( "Estimated value of sigma2:\n ")
> print( corti.nl7$sigma2); cat( "\n\n")
```

Plotting the observed and fitted values of the response and the residuals when the regression the function is a asymmetric sigmoidally shaped curve and the variance is heteroscedastic

We plot the observed and fitted values of the response and the residuals:

```
>      # Plot the observed and fitted values of the response
>      # against the log-dose
> plot(logdose,corti$cpm,xlab="log-dose",ylab="response")
> title(main="Cortisol example",
         sub="Observed and adjusted response")
> lines(unique(logdose),corti.nl7$response)
> plres(corti.nl7,
         wanted=list(F.R=T),
         absolute=T,st=T,smooth=T,
         title="Cortisol example",ask.modify=T)
> abline(0,0)
```

(See figure 4.9, page 102).

Trajectory of roots of maize: how to detect correlations in errors

This is a new example, just introduced in this chapter. We first create a *data-frame* to store the experimental data. Then we plot the observed values of the response against the independent variable.

Creating the data

Experimental data are stored in a *data-frame* called `root`:

```
> root_data.frame(depth = c(0, 0.687, 1.523, 1.981, 2.479,
          3.535, 4.46, 5.506, 6.582, 7.519, 8.615, 9.591,
          10.588, 11.604, 12.471, 13.507,
14.534, 15.45, 16.496, 17.443, 18.5, 19.516, 20.603,
          21.53, 22.547,  23.514, 24.51, 25.606, 26.503, 27.579,
          28.476, 29.512, 30.569, 31.675, 32.613, 33.539, 34.626,
          35.563, 36.599, 37.636, 38.712, 40.068, 40.657,
41.445, 41.983, 42.94, 43.897, 44.934, 45.941, 46.948,
          47.935, 48.982, 50.049, 50.996, 51.913, 52.95, 53.887,
          54.904, 56.021, 57.028, 57.975, 59.002, 59.979, 61.046,
          61.963, 63.079, 63.937, 65.113, 66.18, 66.958,
67.965, 68.942, 70.098, 70.985, 71.952, 72.849, 74.165,
          75.082, 75.88,  77.007, 78.054, 78.174, 79.091, 80.128,
          81.035, 82.032, 83.019, 84.185, 85.042, 86.089, 87.156,
          88.023, 89.15, 89.997, 91.124, 92.829),
dist =
c(-0.009, 3.012, 6.064, 8.726, 10.102, 12.166, 19.076,
          21.967, 23.882, 25.916, 27.143, 28.997, 31.101, 32.667,
          34.691, 36.227, 37.643, 39.827, 41.951, 43.227, 44.105,
          44.584, 45.053, 45.691, 46.25, 47.038, 49.561,  51.077,
          52.523, 55.005, 55.833, 56.432, 57.1, 57.749, 58.178,
```

```
                58.966, 59.823, 61.708, 62.476, 62.945, 63.873, 63.763,
                63.789, 63.889, 64.188, 64.348, 64.598, 64.628, 64.638,
                64.678, 64.559, 64.47, 64.4, 64.271, 64.042, 63.943,
                63.873, 63.844, 63.983, 64.253, 64.323, 64.333, 64.354,
64.643, 64.933, 65.122, 65.222, 65.372, 65.522, 65.523,
                65.523, 65.533, 65.932, 66.361, 66.691, 67.04, 67.459,
                67.549, 67.579, 67.59, 67.161, 65.825, 65.477, 65.886,
                66.415, 66.844, 67.233, 67.901, 67.911, 68.251,
68.251, 68.241, 68.621, 68.92, 68.92, 68.383))
```

Plotting the observed responses against the independent variable

To plot the observed responses versus the depth in the ground, we use the
graphical function `pldnls2`:

```
> pldnls2(root,response.name="dist",X.names="depth")
```

(See figure 4.10, page 103).

Parameter estimation

We describe the model defined on page 103 in a file called `root.mod1`:

```
resp dist;
varind depth;
parresp b,g;
subroutine;
begin
dist=b*(1-exp(-g*depth));
end
```

To estimate the parameters, we use the function `nls2`:

```
> root.nl1<-nls2(root,"root.mod1",c(70,1))
>     # Print the estimated values
> cat( "Estimated values of the parameters:\n ")
> print(root.nl1$theta); cat( "\n\n")
```

Plotting the observed and fitted values of the response and the residuals

We plot the observed and fitted responses versus the depth in the ground
by using the function `plfit`. Option `wanted` specifies the graph requested:
`X.OF` means that observed and fitted values are plotted against the independent variable. Then we plot the residuals by using the function `plres`.
Here, with option `wanted`, we ask for the plot of the residuals versus the
preceding residual (see Figure 4.11, page 104):

```
>     # Plot the observed and fitted values of the response
>     # against the independent variable.
> plfit(root.nl1,
          wanted=list(X.OF=T),
          title="Trajectory of roots")
>     # Plot the residuals
> plres(root.nl1,
```

```
wanted=list(R.R=T),
title="Trajectory of roots")
```

(Results for this example are given in section 4.2.5, page 103.)

Simulated example

Creating the first data-set

The way data are generated is fully explained on page 105.

To generate the 10 values of ε_i, $i = 1, \dots 10$, we use the **S-PLUS** function rnorm (random generation for the normal distribution, with given values of means and standard deviations):

```
# x1<-seq(0.1,1.,length=10)
# exserr<-rnorm(10, mean=0, sd=.05)
# exsw1<-data.frame(x=x1,
#                        y=1-exp(-x1)+exserr)
```

Actually, we do not execute these commands because random generation would not create the same data as shown page 105. Instead, we create the *data-frame* explicitly:

```
> exsw1 <- data.frame(
    x = c(0.1, 0.2, 0.3, 0.4, 0.5, 0.6, 0.7, 0.8, 0.9, 1),
    y = c(0.1913, 0.0737, 0.2702, 0.4270, 0.2968, 0.4474,
            0.4941, 0.5682, 0.5630, 0.6636))
```

Parameter estimation with the first data-set

We describe the model defined on page 105 in a file called sim.mod1:

```
resp y;
varind x;
parresp p1,p2;
subroutine;
begin
y=p1*(1-exp(-p2*x));
end
```

The function nls2 is used to estimate the parameters:

```
> sim.nl1<-nls2(exsw1,"sim.mod1",c(1,1))
>     # Print the main results
> cat( "Estimated values of the parameters:\n ")
> print(coef(sim.nl1))
> cat( "Estimated value of sigma2:\n ")
> print(sim.nl1$sigma2); cat( "\n\n")
```

We calculate the confidence intervals for the parameters:

```
> confsim <- confidence(sim.nl1)
> cat( "Confidence interval for the parameters:\n")
> print(confsim$normal.conf.int); cat( "\n\n")
```

Plotting the generated and fitted values of the response with the first data-set

We plot the generated and fitted values of the response versus x by using the function plfit:

```
> plfit(sim.nl1, wanted=list(X.OF=T),title="Simulated example 1")
```

(See figure 4.12, page 106).

Plotting sensitivity functions for the first data-set

The sensitivity functions Φ_1 and Φ_2 described on page 106 are calculated in the nls2.object for the values of x in the data-set. We plot their values versus x:

```
> matplot(exsw1$x, sim.nl1$d.resp,type="l",xlab="x",ylab="")
> legend(x=0.5,y=.3,legend=c("phi1","phi2"),lty=c(1,2))
> title(" Plot of sensitivity functions")
```

(See figure 4.13, page 107).

Creating the second data-set

Another data set is generated with more appropriate values of x (see page 106):

```
# exsw2 <- data.frame(x=seq(0.1,4.6,length=10),
#                       y=1-exp(-x2)+exserr)
```

To consider the same data as shown on page 106, we do not execute these commands but type the values explicitly:

```
> exsw2  <-  data.frame(
        x = c(0.1, 0.6, 1.1, 1.6, 2.1, 2.6, 3.1, 3.6, 4.1, 4.6),
        y = c(0.1913, 0.3436, 0.6782, 0.8954, 0.7809, 0.9220,
              0.9457, 0.9902, 0.9530, 1.0215))
```

Parameter estimation with the second data-set

The parameters are estimated by using nls2:

```
>   sim.nl2 <- nls2(exsw2,"sim.mod1",c(1,1))
>     # Print the main results
> cat( "Estimated values of the parameters:\n ")
> print(coef(sim.nl2))
```

Plotting the generated and fitted values of the response with the second data-set

We plot the generated and fitted values of the response responses versus x by using the function plfit (see figure 4.14, page 107):

```
> plfit(sim.nl2,
        wanted=list(X.OF=T),
        title="Simulated example 2")
```

(Results for this example are given in section 4.2.6, page 104.)

Cortisol assay example: misspecification tests

Comparison of nested models

The test of the hypothesis H against the alternative A defined on page 108 is done by calculating the difference of the log-likelihoods stored in `corti.nl6` and `corti.nl7`. We compare it to the 0.95 quantile of a χ^2 with 1 degree of freedom (see section 4.3.1, page 108):

```
> Sl <- 64*(corti.nl6$loglik - corti.nl7$loglik)
> cat( "Sl: ", Sl,  "X2(0.95,1):  ",qchisq(0.95,1),"\n ")
```

Test when the regression function is symmetric and the variance proportional to the squared expectation

The results of estimation with the first model, H, defined in section 4.3.3, page 110 (the regression function is symmetric and the variance proportional to the squared expectation), have been stored in the structure called `corti.nl6` (see page 121).

We test the model H against the alternative A_2 (homogeneous variance) by calculating the test statistic S_L. We compare it to the 0.95 quantile of a χ^2 with 25 degrees of freedom (see section 4.3.2, page 108):

```
> Sl <- sum(corti.nl6$replications*log(corti.nl6$variance))-
        sum(corti.nl6$replications*log(corti.nl6$data.stat$S2))
> k <- length(corti.nl6$replications) # number of observations
> cat( "Sl:", Sl,
  "\nX2(0.95,25):",qchisq(0.95, 2*k - 5),"\n\n"  )
```

Test when the regression function is an asymmetric sigmoidally shaped curve and the variance heteroscedastic

The results of the estimation when the regression function is an asymmetric sigmoidally shaped curve and the variance heteroscedastic (the second model defined in section 4.3.3, page 110) have been stored in the structure `corti.nl7` (see page 122).

We calculate the test statistic S_L and compare it to the 0.95 quantile of a χ^2 with 24 degrees of freedom:

```
> Sl <- sum(corti.nl7$replications*log(corti.nl7$variance))-
        sum(corti.nl7$replications*log(corti.nl7$data.stat$S2))
> cat( "Sl:", Sl,
  "\nX2(0.95,24):",qchisq(0.95, 2*k - 6),"\n\n"  )
```

(Results for this example are given in section 4.3.3, page 110.)

Ovocytes example: graphic and misspecification tests using replications

The results of the estimation for the model defined on page 110, in section 4.3.4, have been stored in the structure ovo.nl1 (see page 25).

Plotting the residuals and the variances

We use the function plres to plot the residuals and the absolute residuals versus the fitted values of the response, and the function plvar to plot the empirical variances versus the empirical means:

```
>    # Plot the residuals
> plres(ovo.nl1, wanted=list(F.R=T), absolute=T,
        title="Ovocytes example")
>    # Plot the variances
> plvar(ovo.nl1,wanted=list(Y.S2=T),smooth=T,
        title="Ovocytes example")
```

(See figure 4.15, page 111, and figure 4.16, page 112).

Test

We calculate the test statistic S_L and compare it to the 0.95 quantile of a χ^2 with 53 degrees of freedom:

```
> n <- sum(ovo.nl1$replications) # total number of replications
> k <- length(ovo.nl1$replications) # number of observations
> S1 <- n*log(ovo.nl1$sigma2) -
        sum(ovo.nl1$replications*log(ovo.nl1$data.stat$S2))
> cat( "S1:", S1,
  "\nX2(0.95,53):",qchisq(0.95, 2*k - 5),"\n\n"  )
```

(Results for this example are given in section 4.3.4, page 110.)

Solubility of peptides example: numerical troubles during the estimation process

Estimation with the model defined by equation (4.6). The model defined by equation (4.6) is described in a file called pept.m13:

```
% model pept.m13
resp solubility;
varind RetTime;
var v;
aux a1;
parresp  ed50, sl ;
parvar   h1, h2;
subroutine ;
begin
a1 = 1 + exp (sl*(RetTime-ed50)) ;
solubility = 100. /a1 ;
```

```
v= 1 + h1*solubility*(100+h2-solubility);
end
```

We estimate the parameters with this model. Here, to be able to draw figure 4.17, page 114 (variations of V versus the parameter values at iterations $250, 255, \ldots 500$), we must request that the results calculated thorough the iterative process be saved. With the `control` argument, we request that the estimated values of the parameters and the statistical criterion are saved after every five iterations. By using the option `wanted.print`, we suppress the printing of intermediary results:

```
> pept.ctx13<-list(theta.start=pept.nl3$theta,
              beta.start=c(pept.nl3$beta,0),max.iters=500)
> pept.ctr13<-nls2.control(
    freq=5,step.iters.sv=1,
     wanted.iters.sv=list(iter=T,estim=T,stat.crit=T),
     wanted.print=list(iter=F,stat.crit=F,stop.crit=F,
                 estim=F,fitted=F,num.res=F,sigma2=F))
> pept.nl13<-nls2(pept.d, "pept.m13", pept.ctx13,control=pept.ctr13)
> # Print the result
> summary(pept.nl13)
```

The intermediary results are saved in the component `iters.sv` of the returned structure. The last 50 elements contain the results obtained at iterations $250, 255, \ldots 500$. So, we can draw figure 4.17, page 114, using the following commands:

```
> ind<-50:100
> par(mfrow=c(2,2))
> plot(pept.nl13$iters.sv$theta[ind,1],
  -75*pept.nl13$iters.sv$stat.crit[ind]/2)
> plot(pept.nl13$iters.sv$theta[ind,2],
  -75*pept.nl13$iters.sv$stat.crit[ind]/2)
> plot(pept.nl13$iters.sv$beta[ind,1],
  -75*pept.nl13$iters.sv$stat.crit[ind]/2)
> plot(pept.nl13$iters.sv$beta[ind,2],
  -75*pept.nl13$iters.sv$stat.crit[ind]/2)
```

Estimation with the model defined by equation (4.7). The model defined by equation (4.7) is described in a file called `pept.m14`:

```
% model pept.m14
resp solubility;
varind RetTime;
var v;
aux a1;
parresp ed50, sl ;
parvar  h1, h2;
subroutine ;
begin
a1 = 1 + exp (sl*(RetTime-ed50)) ;
solubility = 100. /a1 ;
v= exp(h1*solubility*(100+h2-solubility));
```

```
end
```

We estimate the parameters with this model:

```
> pept.ctx14<-list(theta.start=pept.nl3$theta,
                   beta.start=c(0.001,0),max.iters=500)
> pept.nl14<-nls2(pept.d, "pept.m14", pept.ctx14)
> # Print the result
> summary(pept.nl14)
```

Peptides example: concluded

We draw plots from the structure `pept.nl14` (see figure 4.18, page 116):

```
>     # Plot the observed and fitted values of the response
>     # against the independent variable.
> plfit(pept.nl14, wanted=list(X.OF=T),
        title="Solubility of peptides - exponential variance",
        sub="Observed and fitted response")
>     # Plot the residuals
> plres.nls2(pept.nl14,  wanted=list(F.R=T),
        smooth=T, ask.modify=T,
        title="Solubility of peptides - exponential variance")
> abline(0,0)
> plres.nls2(pept.nl14,  wanted=list(F.R=T),
        absolute=T, st=T, smooth=T, ask.modify=T,
        title="Solubility of peptides - exponential variance")
```

For both models, we plot the squared residuals versus the fitted values of the regression function. We add a line to join the fitted values of the variances:

```
> par(mfrow=c(1,2))
> plot(range(pept.nl14$response),
       range(c(pept.nl14$residuals**2,pept.nl14$variance)),
       xlab="fitted responses", ylab="squared residuals")
> title(main="Parabolic variance function")
> points(rep(pept.nl3$response,pept.nl3$replications),
       pept.nl3$residuals**2)
> lines(sort(pept.nl3$response),
       pept.nl3$variance[order(pept.nl3$response)],lty=1)
> plot(range(pept.nl14$response),
       range(c(pept.nl14$residuals**2,pept.nl14$variance)),
       xlab="fitted responses", ylab="squared residuals")
> title(main="Exponential variance function")
> points(rep(pept.nl14$response,pept.nl14$replications),
       pept.nl14$residuals**2)
> lines(sort(pept.nl14$response),
       pept.nl14$variance[order(pept.nl14$response)],lty=1)
```

(See figure 4.19, page 117).

5

Calibration and Prediction

In this chapter, we describe how to calculate prediction and calibration confidence intervals. These intervals account for the double source of variability that exists in this type of experiment: the variability of the response about its mean, and the uncertainty about the regression parameters.

Here we provide both bootstrap and asymptotic methods first for the prediction confidence intervals and then for calibration confidence intervals with constant variance. We also demonstrate how to construct calibration confidence intervals when the variances are not constant.

5.1 Examples

To illustrate the prediction problem, let us recall the pasture regrowth example and assume that we are interested in predicting the yield y_0 at the time $x_0 = 50$. We wish to obtain point and interval predictors for the response y_0, which has not been observed.

In example 1.1.2, we estimated the calibration curve of an RIA of cortisol. The real interest of the assay experiment, however, lies in estimating an unknown dose of hormone contained in a new preparation. For this new preparation, we observe a response y_0, and we want to draw inferences about the true dose x_0 corresponding to the new value y_0 of the count. This is a typical calibration problem.

To further illustrate the problem of calibration, let us introduce another example from a bioassay described by Racine-Poon [RP88].

TABLE 5.1. Data from the calibration experiment on nasturtium

Concentration in g/ha	Weight in mg					
.000	920	889	866	930	992	1017
.025	919	878	882	854	851	850
.075	870	825	953	834	810	875
.250	880	834	795	837	834	810
.750	693	690	722	738	563	591
2.000	429	395	435	412	273	257
4.000	200	244	209	225	128	221

TABLE 5.2. Observed weights corresponding to new soil samples from the field

New experiment	Weight in mg		
x_0?	309	296	419

Bioassay on nasturtium

The objective of this experiment is to determine the concentrations of an agrochemical present in soil samples. To this end, bioassays are performed on test plants, a type of cress called *nasturtium*.

In a first step, called the calibration experiment, six replicates Y_{ij} of the response are measured at seven predetermined concentrations x_i of the soil sample. The response is the weight of the plant after 3 weeks' growth, as shown in table 5.1. In a second step, three replicates of the response at an unknown concentration of interest x_0 are measured; they are listed in table 5.2. We would like to estimate x_0. Figure 5.1 shows the responses observed in the calibration experiment as a function of the logarithm of the concentration. The 0 concentration is represented in figure 5.1 by the value -5. A logit-log model is used to describe the concentration response relationship:

$$
\begin{aligned}
f(x, \theta) &= \theta_1, & x = 0 \\
&= \theta_1/(1 + \exp(\theta_2 + \theta_3 \log(x))), & x > 0,
\end{aligned}
$$

where x is the concentration and $f(x, \theta)$ is the expected weight of nasturtium at concentration x.

The model under study is the following one:

$$
\left.
\begin{aligned}
Y_{ij} &= f(x_i, \theta) + \varepsilon_{ij} \\
\mathrm{Var}(\varepsilon_{ij}) &= \sigma^2
\end{aligned}
\right\}, \tag{5.1}
$$

where $j = 1, \ldots 6$, $i = 1, \ldots 7$ and the ε_{ij} are independent centered random variables.

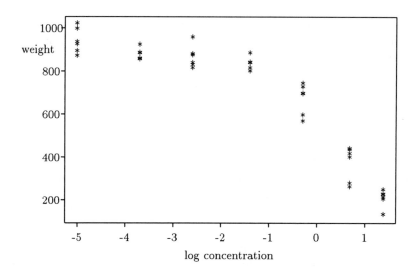

FIGURE 5.1. Nasturtium example: observed weights versus the log concentration

5.2   Problem formulation

In prediction or calibration experiments, we first observe a training or calibration data set, (x_i, Y_{ij}), $j = 1, \ldots, n_i$, $n = \sum n_i$, from which the response curve (or calibration curve) and the variance structure are fitted. Let $\widehat{\theta}$ be the least squares estimator of θ in the case of homogeneous variances $(\mathrm{Var}(\varepsilon_{ij}) = \sigma^2)$, and let it be the maximum likelihood estimator in the case of heterogeneous variances (see chapter 3).

The real interest lies in the new pair (x_0, y_0). In prediction problems, x_0 is given and we want to predict the associated response y_0. The standard estimate for y_0 is $f(x_0, \widehat{\theta})$. In calibration problems, y_0 is observed but the unknown value x_0 must be estimated. The usual estimate of x_0 is the set of all x values for which $f(x_0, \widehat{\theta}) = y_0$. For f strictly increasing or decreasing, the estimate will be a single value obtained by inverting the regression function : $\widehat{x_0} = f^{-1}(y_0, \widehat{\theta})$.

5.3   Confidence intervals

5.3.1   Prediction of a response

Let us begin by discussing the problem of prediction with constant variances, which will provide motivation for the more prevalent problem of calibration.

Given the value x_0, the variance of the error made in predicting y_0 is $\mathrm{Var}(y_0 - f(x_0, \widehat{\theta}))$. If homogeneous variances are assumed, we first obtain

$\widehat{\theta}$. Then we estimate σ^2 by the residual sum of squares $\widehat{\sigma}^2$ and the variance of $f(x_0, \widehat{\theta})$ by $\widehat{S}$, as explained in section 2.3.1. The required prediction interval is

$$\widehat{I}_N(x_0) = \left[f(x_0, \widehat{\theta}) - \nu_{1-\alpha/2}\widehat{S}_{\widehat{f}} \,;\, f(x_0, \widehat{\theta}) + \nu_{1-\alpha/2}\widehat{S}_{\widehat{f}} \right], \qquad (5.2)$$

where

$$\widehat{S}_{\widehat{f}}^2 = \widehat{\sigma}^2 + \widehat{S}^2, \qquad (5.3)$$

and where ν_α is the α-percentile of a variate distributed as a $\mathcal{N}(0,1)$. Under the assumption that y_0 is a Gaussian variable, $\widehat{I}_N(x_0)$ has asymptotic level $1-\alpha$. It is interesting to compare equation (5.2) with equation (2.2), taking $\lambda = f(x_0, \theta)$, because equation (2.2) provides a confidence interval for the mean $f(x_0, \theta)$ rather than for the response itself.

Remark. If the size n of the calibration data set is large, the variance term is dominated by $\widehat{\sigma}^2$, signifying that the uncertainty in predicting y_0 is mostly due to its variability about its mean. The second term, $\widehat{S}^2$, is the correction for estimation of θ. However, for moderate n, $\widehat{S}^2$ can be of the same order of magnitude as $\widehat{\sigma}^2$.

Bootstrap prediction intervals

The bootstrap method described in section 2.3.5 replaces the normal percentiles ν_α, $\nu_{1-\alpha/2}$ in equation (5.2) by bootstrap percentiles using the bootstrap distribution as a better approximation to the distribution of

$$\widehat{T} = \frac{y_0 - f(x_0, \widehat{\theta})}{\left\{ \widehat{\sigma}^2 + \widehat{S}^2 \right\}^{1/2}}.$$

As a result, we obtain more accurate prediction intervals.

Bootstrap prediction intervals are based on the quantiles of

$$\widehat{T}^\star = \frac{y_0^\star - f(x_0, \widehat{\theta}^\star)}{\left\{ \widehat{\sigma}^{2\star} + S_{\widehat{\theta}^\star}^2 \right\}^{1/2}},$$

where $\widehat{\theta}^\star$ and $\widehat{\sigma}^{2\star}$ are bootstrap estimations of θ and σ^2 that are calculated from bootstrap samples $(x_i, Y_{ij}^\star)$, $j = 1, \ldots, n_i$, $i = 1, \ldots, k$, where

$$Y_{ij}^\star = f(x_i, \widehat{\theta}) + \varepsilon_{ij}^\star,$$

and where

$$y_0^\star = f(x_0, \widehat{\theta}) + \varepsilon_0^\star$$

is the "new" bootstrap observation. Here, $(\varepsilon_{ij}^\star, \varepsilon_0^\star)$ is an independent bootstrap sample of the errors of length $n + 1$. The method of generating the errors ε_{ij}^{*} and ε_{0}^{*} from the residuals $\widehat{\varepsilon}_{ij} = Y_{ij} - f(x_i, \widehat{\theta})$ is detailed in section 2.3.5.

Let B be the number of bootstrap simulations. $(\widehat{T}^{*,b}, b = 1, \ldots B)$ is a B-sample of $\widehat{T}^{*}$. Let b_α be the α-percentile of the $\widehat{T}^{*,b}$ (the method of calculating b_α is detailed in section 2.4.1). This gives a bootstrap prediction interval for y_0 of

$$\widehat{I}_B(x_0) = \left[f(x_0, \widehat{\theta}) - b_{1-\alpha/2}\widehat{S}_{\widehat{f}} \; ; f(x_0, \widehat{\theta}) - b_{\alpha/2}\widehat{S}_{\widehat{f}} \right], \qquad (5.4)$$

where $\widehat{S}_{\widehat{f}}$ is defined by equation (5.3).

For large n and B the coverage probability of $\widehat{I}_B(x_0)$ is close to $1 - \alpha$. In practice however, a value of B around 200 usually suffices.

5.3.2   Calibration with constant variances

To do a calibration procedure, suppose now that we observe m replicates y_{0l}, $l = 1, \ldots, m$ of the response at an unknown x_0 and that we wish to obtain a calibration interval for x_0. Let $\overline{y_0} = \sum_{l=1}^{m} y_{0l}/m$. We suppose that

$$\mathrm{Var}(y_{0l}) = \sigma^2 = \mathrm{Var}(Y_{ij}).$$

A calibration interval can be constructed for x_0 by inverting equation (5.2). In other words, the confidence interval is the set of all x such that $\overline{y_0}$ falls in the corresponding prediction interval $\widehat{I}(x_0)$. A calibration interval for x_0 is then

$$\widehat{J}_{\mathcal{N}} = \left\{ x, \quad |\overline{y_0} - f(x, \widehat{\theta})| \leq \nu_{1-\alpha/2} \left\{ \frac{\widehat{\sigma}^2}{m} + \widehat{S}^2 \right\}^{1/2} \right\}, \qquad (5.5)$$

where $\widehat{S}^2$ is the estimation of the asymptotic variance of $f(x_0, \widehat{\theta})$ calculated at $\widehat{x}_0$.

Like the confidence intervals constructed in section 2.3.2, $\widehat{J}_{\mathcal{N}}$ has asymptotic level $1 - \alpha$: Its coverage probability should be close to $1 - \alpha$ when both the number of replications m and the size of the calibration data set n tend to infinity. From our experience, however, $\widehat{J}_{\mathcal{N}}$ may have a covering probability that is significantly different from the expected asymptotic level when n and m have small values.

Bootstrap calibration intervals

By using the bootstrap method presented in section 2.3.5, we can construct calibration intervals that actually achieve the desired nominal level, even in small sample situations.

The construction of bootstrap calibration intervals is based on the simulation of bootstrap samples of the training data (x_i, Y_{ij}^{*}), $j = 1, \ldots, n_i$, $i = 1, \ldots k$ and on bootstrap samples y_{0l}^{*} of the responses at x_0. The bootstrap procedure is as follows:

- Let $\tilde{\varepsilon}_{ij}$ be the n centered residuals:

$$\tilde{\varepsilon}_{ij} = Y_{ij} - f(x_i, \widehat{\theta}) - n^{-1} \sum_{i,j} \left(Y_{ij} - f(x_i, \widehat{\theta}) \right),$$

$j = 1, \ldots, n_i$, $i = 1, \ldots k$. n bootstrap errors ε_{ij}^* are drawn with replacement from the $\tilde{\varepsilon}_{ij}$, each with probability $1/n$. Then $Y_{ij}^* = f(x_i, \widehat{\theta}) + \varepsilon_{ij}^*$. Let $\widehat{\theta}^*$ and $\widehat{\sigma}^{2*}$ denote the bootstrap estimates obtained from the bootstrap sample (x_i, Y_{ij}^*).

- To resample the responses y_{0l}, $l = 1, \ldots, m$, randomly choose m additional bootstrap errors e_l^* from the $\tilde{\varepsilon}_{ij}$ and compute the bootstrap responses at x_0 as $y_{0l}^* = \overline{y_0} + e_l^*$ $l = 1, \ldots, m$. Let $\overline{y_0}^* = \sum_1^m y_{0l}^*/m$.

The resampling scheme can be repeated B times. At each run, we compute

$$\widehat{T}^* = \frac{\overline{y_0}^* - f(\widehat{x_0}, \widehat{\theta}^*)}{\left\{ \frac{1}{m} \widehat{\sigma}^{2*} + \widehat{S}^{2*} \right\}^{1/2}},$$

where $\widehat{x_0} = f^{-1}(\overline{y_0}, \widehat{\theta})$ and $\widehat{S}^* = S_{\widehat{\theta}^*}$ is calculated at the x value $\widehat{x}^* = f^{-1}(\overline{y_0}^*, \widehat{\theta}^*)$. As a result, we obtain B values $\widehat{T}^{*,b}$, $b = 1, \ldots B$ of the bootstrap statistic. Let b_α be the α-percentile of the $\widehat{T}^{*,b}$ (the method of calculating b_α is detailed in section 2.4.1). This gives a bootstrap calibration interval for x_0 of

$$\widehat{J}_B = \left\{ x, \quad b_{\alpha/2} \widehat{S}_{\widehat{f}} \le \overline{y_0} - f(x, \widehat{\theta}) \le b_{1-\alpha/2} \widehat{S}_{\widehat{f}} \right\}, \tag{5.6}$$

where $\widehat{S}_{\widehat{f}}$ is defined in equation (5.3). If, for instance, f is strictly increasing, we can explicitly compute the bounds of the interval:

$$\widehat{J}_B = \left[f^{-1} \left(\overline{y_0} - b_{1-\alpha/2} \widehat{S}_{\widehat{f}}, \theta \right), \quad f^{-1} \left(\overline{y_0} - b_{\alpha/2} \widehat{S}_{\widehat{f}}, \theta \right) \right].$$

From the asymptotic theory, we know that the coverage probability of $\widehat{J}_B$ is close to $1 - \alpha$ for large n and B. In practice, with small sample sizes and B around 200 for instance, $\widehat{J}_B$ has proved to behave reasonably well.

Likelihood ratio type calibration intervals

An alternative to bootstrap intervals is likelihood ratio type intervals. These intervals are based on likelihood ratio test statistics (see section 2.3.3).

Recall that $\widehat{\theta}$ is the least squares estimator of θ, that is, it minimizes the sum of squares

$$C(\theta) = \sum_{i=1}^{k} \sum_{j=1}^{n_i} (Y_{ij} - f(x_i, \theta))^2.$$

The idea is to compare $C(\widehat{\theta})$ with the value of the sum of squares when θ is estimated subject to a known value of x_0. To define the confidence interval, assume that x_0 is known and compute the value of the parameter θ that minimizes the sum of squares defined by

$$\widetilde{C}(\theta) = \sum_{i=1}^{k} \sum_{j=1}^{n_i} (Y_{ij} - f(x_i, \theta))^2 + \sum_{l=1}^{m} (y_{0l} - f(x_0, \theta))^2.$$

Let $\widetilde{\theta}$ denote this value. With these notations, the distribution of the statistic

$$\mathcal{S}_{\mathrm{L}} = (n + m) \left\{ \log \widetilde{C}(\widetilde{\theta}) - \log \left(C(\widehat{\theta}) + \sum_{l=1}^{m} (y_{0l} - \overline{y_0})^2 \right) \right\}$$

can be approximated by a χ^2 with 1 degree of freedom. Equivalently, we can compute the signed root of $\mathcal{S}_{\mathrm{L}}$ defined as

$$R_L(x_0) = \mathrm{sign} \left(\overline{y_0} - f(x_0, \widehat{\theta}) \right) \sqrt{\mathcal{S}_{\mathrm{L}}}.$$

The limiting distribution of R_L is a $\mathcal{N}(0, 1)$ distribution. We can deduce a confidence region for x_0 with asymptotic level $1 - \alpha$:

$$\widehat{J}_R = \left\{ x, \quad \nu_{\alpha/2} \leq R_L(x) \leq \nu_{1-\alpha/2} \right\}. \tag{5.7}$$

We use some numerical computations to obtain the endpoints of $\widehat{J}_R$. Note that $\widehat{J}_R$ is not necessarily an interval, but in practical studies it turns out to be one generally.

Like $\widehat{J}_B$, $\widehat{J}_R$ has proved to perform well in practice. Its coverage probability is close to the desired confidence level.

5.3.3 Calibration with nonconstant variances

We have seen in chapter 3 that, in the case of heterogeneous variances, the appropriate estimation method is the maximum likelihood method. Recall that the log-likelihood based on the observations Y_{ij} is

$$\log L_n \left(Y_{11}, \ldots, Y_{kn_k}, \theta, \sigma^2, \tau \right) =$$
$$-\frac{n}{2} \log 2\pi - \frac{1}{2} \sum_{i=1}^{k} \left[n_i \log \left(g(x_i, \sigma^2, \theta, \tau) \right) + \frac{\sum_{j=1}^{n_i} (Y_{ij} - f(x_i, \theta))^2}{g(x_i, \sigma^2, \theta, \tau)} \right].$$

Adding into the log-likelihood the observations y_{0l}, $l = 1, \ldots, m$, we obtain the global log-likelihood corresponding to our calibration model

$$V \left(\theta, \sigma^2, \tau, x_0 \right) = \log L_n \left(Y_{11}, \ldots, Y_{kn_k}, \theta, \sigma^2, \tau \right)$$
$$-\frac{m}{2} \log 2\pi - \frac{m}{2} \log \left(g(x_0, \sigma^2, \theta, \tau) \right) - \frac{1}{2} \frac{\sum_{l=1}^{m} (y_{0l} - f(x_0, \theta))^2}{g(x_0, \sigma^2, \theta, \tau)}.$$

Let $\widehat{\theta}$, $\widehat{\sigma}^2$, $\widehat{\tau}$, $\widehat{x_0}$ be the maximum likelihood estimators. A calibration interval for x_0 based on the asymptotic distribution of $\overline{y_0} - f(x_0, \widehat{\theta})$ is then just $\widehat{J}_{\mathcal{N}}$.

We can also, and this is what we recommend, compute a calibration interval based on the log-likelihood ratio (see section 3.4.4). That is, we construct an analogous calibration interval as $\widehat{J}_R$ in the case of non-constant variances. Compute $\widetilde{\theta}$, $\widetilde{\sigma}^2$, $\widetilde{\tau}$ that maximize $V(\theta, \sigma^2, \tau, x_0)$ given a fixed value of x_0. With these notations, the distribution of the statistic

$$R_L(x_0) = \text{sign}\left(\overline{y_0} - f(x_0, \widehat{\theta})\right) \times$$
$$\left\{2V\left(\widehat{\theta}, \widehat{\sigma}^2, \widehat{\tau}, \widehat{x_0}\right) - 2V\left(\widetilde{\theta}, \widetilde{\sigma}^2, \widetilde{\tau}, x_0\right)\right\}^{1/2}$$

can be approximated by a $\mathcal{N}(0,1)$ distribution. We can find a confidence interval for x_0 with asymptotic level $1 - \alpha$:

$$\widehat{J}_R = \left\{x, \quad \nu_{\alpha/2} \leq R_L(x) \leq \nu_{1-\alpha/2}\right\}. \tag{5.8}$$

Remark. Currently we cannot propose bootstrap calibration intervals in the case of heterogeneous variances; as already mentioned, bootstrap procedures in an heteroscedastic setting are still under investigation.

5.4   Applications

5.4.1   *Pasture regrowth example: prediction of the yield at time $x_0 = 50$*

Model. The regression function used to analyze the pasture regrowth example (see section 1.1.1) is

$$f(x, \theta) = \theta_1 - \theta_2 \exp\left(-\exp(\theta_3 + \theta_4 \log x)\right),$$

and the variances are homogeneous: $\text{Var}(\varepsilon_i) = \sigma^2$.

Results. The adjusted response curve is

$$f(x, \widehat{\theta}) = 69.95 - 61.68 \exp\left(-\exp(-9.209 + 2.378 \log x)\right).$$

Calculation of prediction intervals for y_0 with asymptotic level 95% (see equation (5.2)).

$\widehat{y_0}$	$\left\{\widehat{\sigma}^2 + \widehat{S}^2\right\}^{1/2}$	$\nu_{0.975}$	$\widehat{I}_{\mathcal{N}}(x_0)$
49.37	1.17	1.96	[47.08 , 51.67]

Calculation of prediction intervals for y_0 with asymptotic level 95%, us-ing bootstrap (see equation (5.4)). The number of bootstrap simulations is $B=199$.

$\widehat{y}_0$	$\left\{\widehat{\sigma}^2 + \widehat{S}^2\right\}^{1/2}$	$b_{0.025}$	$b_{0.975}$	$\widehat{I}_B(x_0)$
49.37	1.17	-1.21	1.65	[47.96 , 51.31]

In this example, the two methods yield the same results: the prediction interval based on normal quantiles and the prediction interval based on bootstrap quantiles are nearly the same. The bootstrap quantiles catch some asymmetry of the distribution of $\widehat{T}$, but this has little effect on the result due to the small values of $\widehat{\sigma}$ and $\widehat{S}$. The numerical difference be-tween the two intervals is due mainly to the variability between bootstrap simulations.

5.4.2   Cortisol assay example

Model. In chapter 4, we concluded that a satisfying model to analyze the cortisol assay data was based on an asymmetric sigmoidally shaped regres-sion function

$$f(x, \theta) = \theta_1 + \frac{\theta_2 - \theta_1}{(1 + \exp(\theta_3 + \theta_4 x))^{\theta_5}}$$

with heteroscedastic variances $\mathrm{Var}(\varepsilon_{ij}) = \sigma^2 f^2(x_i, \theta)$.

Method. The parameters are estimated by maximizing the log-likelihood; see section 3.3.

Results. The maximum likelihood estimators of the parameters are

Parameters	Estimated values
θ_1	133.42
θ_2	2758.7
θ_3	3.2011
θ_4	3.2619
θ_5	0.6084
σ^2	0.0008689

The calibration curve describes the relationship between the dose $d = 10^x$ and the expected response:

$$f\left(d, \widehat{\theta}\right) = \frac{2758.7 - 133.4}{(1 + \exp(3.2011 + 3.2619 \log_{10}(d)))^{0.6084}}.$$

The parameter of interest is the unknown dose of hormone d correspond-ing to the responses obtained from the following new experiment:

New experiment	Response in c.p.m.			
d?	2144	2187	2325	2330

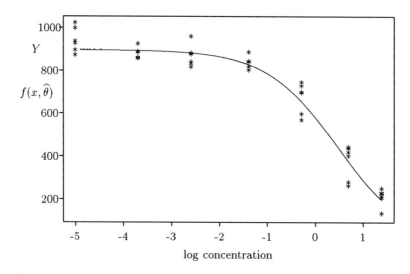

FIGURE 5.2. Nasturtium example: graph of observed and adjusted response values

Since we have nonconstant variances, we compute a likelihood ratio type calibration interval and apply the results of section 5.3.3 with $x_0 = d$.

Calculation of a calibration interval for d, with asymptotic level 95% (see equation (5.8)).

$$\begin{array}{c|c} \widehat{d} & \widehat{J}_R \\ 0.0574 & [0.0516\,,\,0.0642] \end{array}$$

$\widehat{J}_R$ takes into account both the variability of the responses observed at dose d and the uncertainty about the calibration curve.

5.4.3   Nasturtium assay example

Model. The regression function is

$$f(x, \theta) = \theta_1/(1 + \exp(\theta_2 + \theta_3 \log(x))),$$

and the variances are homogeneous: $\mathrm{Var}(\varepsilon_i) = \sigma^2$.

Method. The parameters are estimated by minimizing the sum of squares, $C(\theta)$; see equation (1.10).

Results. The calibration curve is defined by the following equation:

$$f\left(x, \widehat{\theta}\right) = \frac{897.86}{1 + \exp(-0.61 + 1.35 \ \log(x))},$$

and is shown in figure 5.2.

Calculation of a calibration interval for the unknown x_0 corresponding to the responses from the new experiment (see table 5.2). In the case of homogeneous variances, we proposed two calibration intervals. The first one, equation (5.6), is based on bootstrap simulations; the second one, equation (5.7), is based on the computation of the likelihood ratio statistic.

Bootstrap interval. In this example, 500 simulations of bootstrap samples (B=500) are necessary to stabilize the values of the bootstrap percentiles. The following calibration interval was obtained from 500 bootstrap simulations:

$\widehat{x_0}$	$b_{0.025}$	$b_{0.975}$	$\widehat{J}_B$
2.26	-1.55	1.50	[1.88 , 2.73]

Likelihood ratio type interval.

$\widehat{x_0}$	$\widehat{J}_R$
2.26	[1.95 , 2.80]

The two methods do not yield the same result. But we have no argument to prefer one calibration interval over the other.

Note that in this example, the computation of the calibration interval $\widehat{J}_N$, according to equation (5.5), yields: [1.80, 2.91], which is wider than $\widehat{J}_B$ or $\widehat{J}_R$. This is not always the case. As already mentioned, $\widehat{J}_B$ or $\widehat{J}_R$ is to be preferred, both on theoretical grounds and from the results of simulation studies.

5.5   References

The theory underlying the construction of bootstrap and likelihood ratio type calibration intervals can be found in [GHJ93] and [GJ94].

5.6   Using **nls2**

This section reproduces the commands and files used in this chapter to analyze the examples using **nls2**.

The commands introduced in preceding "Using **nls2**" sections are assumed to have already been executed.

Pasture regrowth example: prediction of the yield at time $x_0 = 50$

The results of estimation with the model defined in section 5.4.1, page 138, have been stored in the structure called **pasture.nl1** (see section 1.6, page 20).

In order to compute prediction intervals, we need to evaluate $f(x_0, \widehat{\theta})$ and $\widehat{S}$. The best way to do this is to compute a confidence interval for $f(x_0, \theta)$ by using the function confidence and obtaining $f(x_0, \widehat{\theta})$ and $\widehat{S}$ as by-products.

Calculation of $f(x_0, \widehat{\theta})$ and $\widehat{S}$

The function of the parameters that we consider is the regression function f at time $x_0 = 50$. It is described in a file called pasture.predict. The value of x_0 is set by the key word pbispsi:

```
psi yield;
ppsi p1, p2, p3, p4;
pbispsi x0;
subroutine;
begin
yield = p1-p2*exp(-exp(p3+p4*log(x0)));
end
```

We use the function confidence to compute the values of $f(x_0, \widehat{\theta})$ and $\widehat{S}$:

```
> pasture.conf.fx0<-confidence(pasture.nl1,
          file="pasture.predict",
          pbispsi=50)
>     # Print the results
> cat("Estimated value of f(x0,theta):", pasture.conf.fx0$psi,"\n" )
> cat("Estimated value of S:", pasture.conf.fx0$std.error,"\n" )
```

Prediction interval $\widehat{I}_{\mathcal{N}}(x_0)$ for y_0, with asymptotic level 95%

From the previous results, we calculate the prediction interval $\widehat{I}_{\mathcal{N}}(x_0)$ for y_0 (see equation (5.2), page 134):

```
> variance.y0 <- pasture.nl1$sigma2 + pasture.conf.fx0$var.psi
> sqrt.var <- sqrt(variance.y0)
> lower.y0<- pasture.conf.fx0$psi + qnorm(0.025)*sqrt.var
> upper.y0<- pasture.conf.fx0$psi + qnorm(0.975)*sqrt.var
```

We display the values of $\widehat{y}_0$, $\sqrt{\widehat{\sigma}^2 + \widehat{S}^2}$, $\nu_{0.975}$ and $\widehat{I}_{\mathcal{N}}(x_0)$:

```
> cat("Estimated value of y0:",pasture.conf.fx0$psi,"\n" )
> cat("Estimated value of the std:", sqrt.var,"\n" )
> cat("nu_(0.975):", qnorm(0.975),"\n" )
> cat("Estimated value of In:",lower.y0, upper.y0,"\n" )
```

Bootstrap prediction intervals for y_0

To calculate a bootstrap prediction interval for y_0, $\widehat{I}_B(x_0)$, according to equation (5.4), page 135, we use the function bootstrap with the following option: method="calib". This function returns the bootstrap percentiles of $\widehat{T}^*$:

```
> pasture.pred.y0<-bootstrap(pasture.nl1,
                             n.loops=500,
                             method="calib",
                             file="pasture.predict",
                             ord=pasture.conf.fx0$psi,
                             pbispsi=50)
> lower.y0<- pasture.conf.fx0$psi +
             pasture.pred.y0$conf.bounds[1] *sqrt(variance.y0)
> upper.y0<- pasture.conf.fx0$psi +
             pasture.pred.y0$conf.bounds[2] *sqrt(variance.y0)
```

We display the values of $\widehat{y_0}$, $b_{0.025}$, $b_{0.975}$ and $\widehat{I}_B(x_0)$:

```
> cat("Estimated value of y0:", pasture.conf.fx0$psi,"\n" )
> cat("b_0.025, b_0.975:",pasture.pred.y0$conf.bounds,"\n" )
> cat("Estimated value of Ib:",lower.y0,upper.y0,"\n" )
```

Note: The bootstrap method generates different numbers at each execution. Thus, results of these commands may vary slightly from those displayed in section 5.4.1.

Cortisol assay example

The results of estimation when the regression function is an asymmetric sigmoidally shaped curve and the variances are heteroscedastic have been stored in the structure `corti.nl7` (see page 122).

Likelihood ratio type calibration interval for x_0, with asymptotic level 95%

We use the function `calib` to calculate the calibration interval $\widehat{J}_R$ defined by equation (5.8), page 138, for the unknown dose d of hormone that corresponds to the response values given in section 5.4.2.

Before calling it, we have to:

1. Describe the inverse $f^{-1}(y, \theta)$ of the regression curve.

   We define it in a file called `corti.finv`:

```
abs dose;
ord f;
paract n,d,a,b,g;
pbisabs minf,pinf;
aux a1,a2,a3;
subroutine;
begin
a1 = (f-n)/(d-n);
a2 = exp(-(1/g)*log(a1));
a3 = (log(a2-1)-a)/b;
dose = if f >= d then minf else
       if f <= n then pinf else
       exp(a3*log(10))
fi   fi;
end
```

2. Generate the programs required by the function `calib`.

The operating system commands `analDer`, `crCalib`, and `crInv` are provided with the system **nls2** to do that:

- `analDer` generates the program that calculates the regression model,

- `crCalib` generates the program that calculates d given fixed values of the regression parameters,

- `crInv` generates the program that calculates the inverse of the regression curve.

```
$ analDer corti.mod6
$ crCalib corti.mod6
$ crInv corti.finv
```

3. Compile the generated files:

```
Splus COMPILE corti.mod6.c INC=$SHOME/library/nls2/include
Splus COMPILE corti.mod6.dc.c INC=$SHOME/library/nls2/include
Splus COMPILE corti.finv.c INC=$SHOME/library/nls2/include
```

The programs required by the function `calib` are loaded into our **S-PLUS** session by `loadnls2`. Then we apply `calib`. Using the option `x.bounds`, we specify a research interval:

```
> loadnls2(model=c("corti.mod6.o","corti.mod6.dc.o"),
        inv="corti.finv.o",
        tomyown=paste(getenv("NLS2_DIR"),"/EquNCalib.o", sep=""))
> corti.calib<-calib(corti.nl7,
        file="corti.finv",
        x.bounds=c(0.02,0.08),
        ord=c(2144,2187,2325,2330))
```

We display the values of $\widehat{d}$ and $\widehat{J}_R$:

```
> cat("Estimated value of d:", corti.calib$x,"\n" )
> cat("Jr:",corti.calib$R.conf.int,"\n" )
```

(Results are given in section 5.4.2, page 139.)

Nasturtium assay example

The nasturtium assay example is a new example, only introduced in this chapter. We first create a *data-frame* to store the experimental data. Then, we estimate the parameters, and, finally, we calculate calibration intervals.

Creating the data

The experimental data (see table 5.1, page 132) are stored in a *data-frame* called `nasturtium`:

```
> x.nas <-c(
  0.000,0.000,0.000,0.000,0.000,0.000,
  0.025,0.025,0.025,0.025,0.025,0.025,
  0.075,0.075,0.075,0.075,0.075,0.075,
  0.250,0.250,0.250,0.250,0.250,0.250,
  0.750,0.750,0.750,0.750,0.750,0.750,
  2.000,2.000,2.000,2.000,2.000,2.000,
  4.000,4.000,4.000,4.000,4.000,4.000)
> y.nas <-c(
  920,889,866,930,992,1017,919,878,882,854,851,850,870,825,953,
  834,810,875,880,834,795,837,834,810,693,690,722,738,563,591,
  429,395,435,412,273,257,200,244,209,225,128,221)
> nasturtium<-data.frame(x=x.nas,y=y.nas)
```

Plot of the observed weights versus the log concentration

We plot the observed weights versus the log concentration. The zero concentration is represented by the value -5:

```
> log.nas<-c(rep(-5,6),log(x.nas[7:42]))
> plot(log.nas,y.nas,
        xlab="log-concentration",ylab="response",
        main="Nasturtium example",sub="Observed response")
```

(See figure 5.1, page 133).

Parameters estimation

We describe the model defined by equation (5.1), page 132, in a file called `nas.mod`:

```
resp y;
varind x;
parresp t1,t2,t3;
subroutine;
begin
y =  if x==0 then t1 else
        t1/(1+exp(t2+t3*log(x)))
             fi;
end
```

To estimate the parameters, we apply the function **nls2**. We plot the observed and fitted responses versus the log concentration by using graphical functions of **S-PLUS**. A line joins the fitted values:

```
> loadnls2() # reload the default programs
> nas.nl<-nls2(nasturtium,"nas.mod",c(900,-.5,1))
> plot(log.nas,y.nas,
        xlab="log-concentration",ylab="response",
```

```
                main="Nasturtium example",
                sub="Observed and adjusted response")
> lines(unique(log.nas),nas.nl$response)
```

(See figure 5.2, page 140).

Bootstrap calibration interval for x_0

The function whose confidence interval is requested, i.e, the regression function f, is described in a file called `nas.psi`. The variable to be calibrated is introduced by the key word `pbispsi`:

```
psi y;
ppsi t1,t2,t3;
pbispsi x0;
subroutine;
begin
y =   t1/( 1 + exp( t2 + t3 * log(x0))  );
end
```

To calculate a confidence interval $\widehat{J_B}$ for x_0 according to equation (5.6), page 136, we first use the function `bootstrap` with the argument `method` set to `calib`, which returns bootstrap percentiles, and then the function `calib`:

```
> nas.boot<-bootstrap(nas.nl,method="calib",file="nas.psi",
                      n.loops=500,
                      ord=c(309, 296, 419),
                      pbispsi=nas.calib$x)
> b0.025<-nas.boot$conf.bounds[1]
> b0.975<-nas.boot$conf.bounds[2]
> boot.calib<-calib(nas.nl,file="nas.finv",ord=c(309, 296, 419),
            conf.bounds=c(as.numeric(b0.025),as.numeric(b0.975)))
```

We display the values of $\widehat{x_0}$, $b_{0.025}$, $b_{0.975}$ and $\widehat{J_B}$:

```
> cat("Estimated value of x0:", boot.calib$x,"\n" )
> cat("b_0.025, b_0.975:",nas.boot$conf.bounds ,"\n" )
> cat("Estimated value of Jb:",boot.calib$S.conf.int,"\n" )
```

Results are displayed in section 5.4.3, page 140. Note that bootstrap simulations generate different numbers at each execution and the results of these commands may be slightly different from those reported in section 5.4.3.

Likelihood ratio type calibration interval for x_0 with asymptotic level 95%

To calculate the likelihood ratio type calibration interval $\widehat{J_R}$ defined by equation (5.7), page 137, for x_0, the unknown concentration that corresponds to the response values given in table 5.2, page 132, we use the function `calib`.

In the cortisol example, additional programs were required to use `calib`. Here, the variances are homogeneous and no extra work is necessary. We only have to describe the inverse of the regression function. We define it in a file called **nas.finv**:

```
abs x;
ord y;
paract t1,t2,t3;
aux a1;
subroutine;
begin
a1 = log( (t1/y) - 1) - t2;
x = if y >= t1 then 0 else
    exp( a1/t3 )
    fi;
end
```

We apply the function `calib` and display the values of $\log \widehat{x_0}$ and $\widehat{J}_R$:

```
> nas.calib<-calib(nas.nl, file = "nas.finv",
      ord=c(309, 296, 419))
>    # Print the results
> cat("Estimated value of x0:",  nas.calib$x,"\n" )
> cat("Jr:", nas.calib$R.conf.int,"\n" )
```

The calibration interval $\widehat{J}_\mathcal{N}$ mentioned in the remark at the end of section 5.4.3 is obtained as follows:

```
> cat("Jn:", nas.calib$S.conf.int,"\n" )
```

References

[BB89]   H. Bunke and O. Bunke. Nonlinear regression, functional relations and robust methods analysis and its applications. In O. Bunke, editor, *Statistical Methods of Model Building*, Volume 2. Wiley, New York, 1989.

[BH94]   A. Bouvier and S. Huet. nls2. non-linear regression by s-plus functions. *Computational Statistics and Data Analysis*, 18:187–190, 1994.

[Bun90]   O. Bunke. Estimating the accuracy of estimators under regression models. Technical Report, Humboldt University, Berlin, 1990.

[BW88]   D.M. Bates and D.G. Watts. *Nonlinear Regression Analysis and Its Applications*. Wiley, New York, 1988.

[Car60]   N.L. Carr. Kinetics of catalytic isomerization of n-pentane. *Industrial and Engineering Chemistry*, 52:391–396, 1960.

[CR88]   R.J. Carroll and D. Ruppert. *Transformation and Weighting in Regression*. Chapman and Hall, London, 1988.

[CY92]   C. Chabanet and M. Yvon. Prediction of peptide retention time in reversed-phase high-performance liquid chromatography. *Journal of Chromatography*, 599:211–225, 1992.

[Fin78]   D.J. Finney. *Statistical Method in Biological Assay*. Griffin, London, 1978.

[FM88]   R. Faivre and J. Masle. Modeling potential growth of tillers in winter wheat. *Acta Œcologica, Œcol. Gener.*, 9:179 –196, 1988.

[Gal87]   A.R. Gallant. *Nonlinear Statistical Models*. Wiley, New York, 1987.

[GHJ93]   M.A. Gruet, S. Huet, and E. Jolivet. Practical use of bootstrap in regression. In W. Härdle and L. Simar, editors, *Computer Intensive Methods in Statistics*, pages 150–166. Physica-Verlag, Heidelberg, 1993.

[GJ94]   M.A. Gruet and E. Jolivet. Calibration with a nonlinear standard curve: how to do it? *Computational Statistics*, 9:249–276, 1994.

[Gla80]   C.A. Glasbey. Nonlinear regression with autoregressive time-series errors. *Biometrics*, 36:135–140, 1980.

[HJM89]   S. Huet, E. Jolivet, and A. Messéan. Some simulations results about confidence intervals and bootstrap methods in nonlinear regression. *Statistics*, 21:369–432, 1989.

[HJM91]   S. Huet, E. Jolivet, and A. Messéan. *La régression non-linéaire: méthodes et applications à la biologie*. INRA, Paris, 1991.

[HLV87]   S. Huet, J. Laporte, and J.F. Vautherot. Statistical methods for the comparison of antibody levels in serums assayed by enzyme linked immuno sorbent assay. *Biométrie–Praximétrie*, 28:61–80, 1987.

[LGR94]   F. Legal, P. Gasqui, and J.P. Renard. Differential osmotic behaviour of mammalian oocytes before and after maturation: a quantitative analysis using goat oocytes as a model. *Cryobiology*, 31:154–170, 1994.

[LM82]   J.D. Lebreton and C. Millier. *Modèles Dynamiques Déterministes en Biologie*. Masson, Paris, 1982.

[Rat83]   D.A. Ratkowsky. *Nonlinear Regression Modeling*. M. Dekker, New York, 1983.

[Rat89]   D.A. Ratkowsky. *Handbook of Nonlinear Regression Models*. M. Dekker, New York, 1989.

[Ros90]   G.J.S. Ross. *Nonlinear Estimation*. Springer–Verlag, New York, 1990.

[RP88]   A. Racine-Poon. A bayesian approach to nonlinear calibration problems. *Journal of the American Statistical Association*, 83:650–656, 1988.

[SW89]   G.A.F. Seber and C.J. Wild. *Nonlinear Regression*. Wiley, New-York, 1989.

[TP90]   F. Tardieu and S. Pellerin. Trajectory of the nodal roots of maize in fields with low mechanical constraints. *Plant and Soil*, 124:39–45, 1990.

[VR94]   W.N. Venables and B.D. Ripley. *Modern Applied Statistics with S-Plus*. Springer–Verlag, New York, 1994.

[WD84]   H. White and I. Domowitz. Nonlinear regression with non independant observations. *Econometrica*, 52:143–161, 1984.

[Wu86]   C.F.J. Wu. Jackknife, bootstrap and other resampling methods in regression analysis (with discussion). *The Annals of Statistics*, 14:1291–1380, 1986.

Index

Springer Series in Statistics

(continued from p. ii)